W9-BFX-429

THE
SUPERHUMAN
MIND

THE
SUPERHUMAN
MIND

FREE THE GENIUS
IN YOUR BRAIN

**Berit Brogaard, PhD,
and Kristian Marlow, MA**

HUDSON
STREET
PRESS

HUDSON STREET PRESS
An imprint of Penguin Random House LLC
375 Hudson Street
New York, New York 10014
penguin.com

LIBRARY OF CONGRESS CATALOGING-IN-PUBLICATION DATA

Brogaard, Berit.
  The superhuman mind : free the genius in your brain / Berit Brogaard, PhD, and Kristian
Marlow, MA.
    pages cm
  Includes bibliographical references and index.
  ISBN 978-1-59463-368-3
  1. Self-actualization (Psychology) 2. Self-perception. I. Marlow, Kristian. II. Title.
  BF697.5.S43B75 2015
  153.9—dc23
                            2014041967

Printed in the United States of America
10 9 8 7 6 5 4 3 2 1

Set in New Caledonia LT Std
Designed by Eve L. Kirch

CONTENTS

FOREWORD

Some say that we use only 10 percent of our brain capacity. After my fifty-plus years of working with persons with savant syndrome, both congenital and acquired, I think that may be an overestimate. This book, with its meticulous and vast account of past and current neuroscience research, delivered in a very reader-friendly fashion, reinforces my impressions.

I met my first savants in 1962 when I started a Children's Unit at a hospital here in Wisconsin. One boy had memorized the bus system of the entire city of Milwaukee. Another young patient, mute and severely disabled, could put together a two-hundred-piece jigsaw puzzle, picture side down, just from the geometric shape of the puzzle pieces. A third boy was a walking almanac of what happened on any day in history (this was BG—Before Google). And a fourth little lad made free throws with unerring accuracy, putting his feet in exactly the same place, holding the ball in exactly the same way, and tossing the ball in exactly the same trajectory, like a pitching machine.

It occurred to me then that this rare condition, in which extraordinary ability stands in stark juxtaposition to severe disability, was speaking volumes about hidden brain potential and brain plasticity. This impression was reinforced immensely when I later began to encounter acquired savants—ordinary persons who suddenly showed extraordinary musical, artistic, or mathematical abilities that had apparently lain dormant until a head injury, stroke, or other damage to the central nervous system forced them to the surface.

I documented many cases of both congenital and acquired savants in my 2010 book, *Islands of Genius: The Bountiful Mind of the Autistic, Acquired, and Sudden Savant.* I wrote that both conditions hint at the "little Rain Man" within us all, but noted that the real challenge is how to tap into that dormant potential without enduring some brain injury or central nervous system catastrophe, and how to accomplish that as nonintrusively as possible.

This book addresses that challenge directly. It provides further convincing evidence of dormant brain potential within us all, and provides some clues as to how to tap into it in readily available, nonintrusive ways.

One way of tapping the potential of our "superbrain," if we choose to, will be through intentional and deliberate effort to apply our existing artistic, mathematical, musical, or other skills to learning new abilities such as calendar calculating, lucid dreaming, or adding to our everyday memory capacity using specific mental imagery techniques. The book suggests that even synesthesia or perfect pitch can be learned to some degree, thus adding those elements to the neurotypical brain's repertoire of abilities. Yet that does not detract from, nor explain, the innate, "factory-installed" instant access to the special skills of savants, both congenital and acquired. In these cases, such skills are not simply "learned." In the savant, prodigy, or genius, "nature" contributes as liberally to extraordinary skills as does "nurture," whether for ten thousand hours or more.

Another key to unlocking our brain potential may come from emerging technical approaches. The "tyranny of the left hemisphere" can be limited, temporarily at least, by using certain nonintrusive external electrodes, which allow right-brain, nondominant abilities to emerge—the same abilities associated with savant syndrome. While the idea of a total division between the left brain and right brain is an oversimplification, the fact is the brain hemispheres do specialize in certain functions, so temporarily inhibiting the left hemisphere "brain boss" allows suppressed right hemisphere abilities and capacities to surface. Some naturally occurring nutritional supplements hold promise in enhancing brain function, as do some existing medications and experimental compounds. Evaluating risks and benefits with all of these is a constant balancing act, but formal efforts with these products are underway, as discussed in the book.

An even less dramatic approach, but still an intentional one, is what I refer to as "rummaging around in the right hemisphere." We are generally a left-brain society, and logical, sequential thinking and language—specialties of the left hemisphere—serve us well. But that comes at the expense of the more creative, artistic, spontaneous, and subconscious capacities of the right hemisphere. As this book demonstrates, one can, because of brain plasticity, cause a "shift to the right" by deliberately visiting those capacities more frequently.

Finally, with all we are learning about dormant brain capacity and plasticity in neurotypical persons, as well as in those with exceptional brain function, the additional beneficiaries will be persons with existing handicaps or disabilities. While for most of us the goal is to find nonintrusive methods of brain enhancement, those with serious central nervous system diseases are already turning to techniques such as deep brain stimulation to treat conditions like Parkinson's disease or refractory depression. Like pacemakers for the heart, similar devices for persons with epilepsy might one day be used to abort seizures by applying to the brain the same principles that are used to treat serious heart arrhythmias. Already, a helmet device containing sensitive electrodes can allow some quadriplegic persons to move a computer cursor by merely thinking about it, and can even stimulate paralyzed muscles to restore motion and mobility.

In short, our efforts to explore, understand, and harness this most complex organ and capacity in the human body—the brain—will not only enhance our everyday functioning but will propel us further than ever toward fully maximizing both the brain and human potential.

To that end, this book provides a trailblazing menu.

Darold A. Treffert, MD

ACKNOWLEDGMENTS

First, we must thank our students and student colleagues, whose feedback and questions sparked the initial idea for the book.

We are forever indebted to Jill Marsal, the best agent in the world, for taking us on board, believing in us, and helping us develop a set of rough ideas into a book. We cannot thank you enough.

Caroline Sutton and Brittney Ross's insightful editing and formidable suggestions have made the completion of this book possible. We cannot imagine a better team of editors. Sincere thanks also go to Catherine Knepper for the countless hours of line editing at the initial stages of the project.

This book could not have been written without the research participants, scientists, and friends who agreed to share their stories: Derek Amato, Amber L. Anastasio, Ralph Dittman, Leigh Erceg, Patrick Fagerberg, Daniel Kish, Tom Jacobson, Katherine MacLean, Mark Nissen, Jason Padgett, Lidell Simpson, Peg Schwartz, Robert Waggoner, and dozens of others who preferred to remain anonymous. Thanks also to the fantastic people at the University of Miami, whose positive attitude and kindness made finishing this book so much easier than it could have been.

We are forever indebted to our amazing families, significant others, close friends, and brainy cats, who have inspired and supported us throughout the writing process.

THE
SUPERHUMAN
MIND

CHAPTER 1

The Hidden Abilities in All of Us

Ordinary, Colorful, and Accidental Superminds

As researchers at the Brogaard Lab for Multisensory Research, in Miami, Kristian and I routinely meet people with extraordinary mental abilities: card counters who can beat the house, self-taught mathematicians, people with face blindness who rely on sound "pings" to recognize faces, blind people who navigate the world using echolocation like bats and whales, memory champions who seem to have hard drives in their heads, and people who suddenly become musical virtuosos or accomplished painters after a blow to the head. Truly an eclectic bunch. While many of them are savants with extraordinary abilities, you'd never know it sitting with them at a café. What's more, not a single one was born with prodigious abilities.

Talk about extraordinary mental abilities might bring to mind a limited number of well-known men (and the occasional woman): Benjamin Franklin, Sir Isaac Newton, Albert Einstein, Andy Warhol, Francis Crick, Richard Feynman, Steve Jobs, Marie Curie, or perhaps an eccentric intellectual or a maladjusted loner—a weirdo, a freak, or maybe even a psychopath.

People with extraordinary mental abilities are sometimes looked upon with suspicion or unease, but the truth is that many of us long to *have* these abilities. We long for extraordinary levels of skill and effortless proficiency—and despair over our losing tickets in the genetic lottery of super-ability and talent. The prevailing assumptions are that people are *born* with extraordinary mental abilities, not *made*; that they operate on a plane inaccessible to the rest of us; and that they're among the blessed few with a free pass excusing them from the drudgery of practice and labor-intensive learning.

Nothing could be further from the truth.

After spending a great deal of time with ordinary people with extraordinary abilities, Kristian and I felt compelled to share their collective story, a story that takes issue with the cult of God-given super-abilities that cannot be taught or acquired. Not one of the people we encountered in the lab whose stories we'll share in this book is a born "super-person." Their pathways to brilliance are as varied as their personalities. But whether it was an injury, an innate brain disorder, an occasion of learned synesthesia, or a mentally "downloaded" algorithm that enabled them to bypass slow, conscious thinking, these gifted individuals gained cognitive access to areas of their brains that normally operate behind closed doors. They acquired the ability to manipulate information in new, ingenious ways or at lightning speed—and they have much to teach us about how we can unlock our own hidden talents and abilities.

In some cases the relevant neural areas in our brains are already completing amazing tasks, but they're doing so below the level of conscious awareness. The parietal cortex, located on the top of the head, contains brain regions involved in completing ordinary mathematical tasks like performing arithmetic and solving algebra equations, but it also houses neurons that can solve extremely complicated mathematical problems faster than a computer. For example, parietal neurons calculate the exact hand aperture and hand position required in order for us to quickly reach out and grasp a rapidly moving object or to strike a key really fast on the keyboard while engaging with intellectually challenging text on the screen. While the parietal brain regions solve incredibly complex mathematical problems on a daily basis, most of us do not have access to the calculations or the solutions to the problems. Nor can we purposely make use of these brain regions to perform other types of calculations or to solve a different set of problems. But in rare cases the brain undergoes a functional or structural reorganization that allows just this type of access—and it's this "behind the velvet ropes" access to specialized brain regions that paves the way for the supermind.

One Key to the Supermind

Many folks believe that only special brains can achieve the ultimate level of extraordinary mental ability and that excellence is an inborn talent bestowed on a select few. But it turns out that all brains have hidden superhuman abilities. We just have to use the right keys to unlock them. One such key is synesthesia. Based on our research, one of our hypotheses is that synesthesia can be the brain's way of opening up areas that we don't normally have conscious access to.

To understand how this works, we'll need to know more about this fascinating condition. Synesthesia is a special way of perceiving the world, involving experiences of connections between seemingly unrelated sensations. For example, the number three may lead to a perception of copper green, the word "kiss" may flood the mouth with the flavor of bread soaked in tomato soup, and the key of C# minor may elicit a bright purple spiral radiating from the center of the visual field.

Having these colliding senses can, in rare cases, be a debilitating condition, such as when all the rainbow's colors brutally penetrate the visual field of a particularly sensitive color-to-sound synesthete. Most synesthetes, however, describe their unusual sensations as pleasant. Some experience the condition as an inner art exhibit or a natural wonder.

One of the best-known forms of color synesthesia is grapheme-to-color synesthesia, in which numbers or letters are seen as colored. But lots of other forms of color synesthesia have been identified, including week-to-color synesthesia, sound-to-color synesthesia, taste-to-color synesthesia, and fear-to-color synesthesia.

There are also lesser-known forms of synesthesia. In a 1913 article in the *Journal of Abnormal Psychology*, neurologist Isador Coriat described a case of "colored pain," which is still considered a rare form of synesthesia. These synesthetes perceive colors as they experience pain. Coriat's subject was an intelligent forty-year-old woman suffering from anxiety, sleepwalking, and headaches. As far back as she could remember she'd see different colors when she felt pain. For her, pain produced clear, distinct colors, and a certain "kind of pain" consistently produced a particular color. "Each type of pain produced its individual and invariable color, for instance: Hollow pain, blue color; sore pain, red color; deep headache, vivid scarlet; superficial headache, white color; shooting neuralgic pain, white color." The woman saw colors as masses with no recognizable shape, except when pain "involved a jagged, longitudinal, or round area, the color stimulated by this particular type of pain had a corresponding geometrical figure."

Artist Carol Steen is a contemporary synesthete who experiences colored pain. Her pain is orange. She reports on a "distinctly unpleasant" experience of synesthesia while at a dental appointment:

> I was at the dentist, and he was drilling. And I don't like the sound of the drill—but the color orange that completely

flooded my vision, I couldn't shut my eyes, because they were already shut! [laughs]

Except that I'm able to use it diagnostically. I had to have a root canal done once (not my favorite game) but you know, sometimes when you have a tooth pain you're not quite sure which tooth it is? He said, "I can't really say that you need a root canal in this tooth." I said, "This tooth is orange; please do it." And he hesitated. I said, "Look. If I'm wrong, this tooth will *never* need a root canal." So he went ahead and he did it.

He said—he poked around a little bit—"This tooth needs a root canal." He said, "It hasn't really become 'ripe' yet, but the nerve is dying." And sure enough, when the nerve was out, and the anesthesia had worn off, there was no more orange. It's like orange is my default color for pain.

Synesthesia gets weirder. One of our research participants, Megan, experiences music by touch. The sound of the piano feels like taps on her face. Strings vibrate in her chest. Waves from brass instruments pass in front of her, sometimes buzzing around her neck. Drums come up from below. She may even feel as if she is being slapped on the face by music. This might sound painful, but it's actually an ecstatic experience for Megan. Intensity increases with volume, but these sensations are never unpleasant, as Megan feels like she is *within* the music, a part of it.

My own experiences with synesthesia actually provided the impetus for my current research. Since I was a child I have had vivid visual images in response to fearful or uncomfortable thoughts. The fear-induced synesthetic images look something like a landscape that's projected out into the world about twenty to thirty centimeters from my eyes. The landscape is bluish green, with spiky mountain peaks—picture a large piece of highly wrinkled paper or cloth and you've got the gist. This "landscape" moves as if it's being shaken by some unseen hand; sometimes I see a rotating movement within it. When the fear is strongest the images are extremely vivid, sometimes completely blocking my visual field and obscuring my surroundings. When the fear or discomfort is less intense, the images are present but transparent, and I can still see my surroundings. Not all of my uncomfortable or fearful thoughts are associated with this sort of phenomenology, but the occurrence of it is a sure sign of uncomfortable or scary thoughts.

When I was a child I used to be deadly scared of the moving, wrinkled landscape projected out in front of my head. I had heard somewhere that you see yourself just before you die—a real out-of-body experience. I later found out that the famous psychoanalyst Sigmund Freud suffered from the same anxiety all of his life. I lived in such constant fear of seeing myself that this caused the extreme images to be regularly projected out into my room. I would close my eyes, but it continued behind my eyelids, although a little less intense. I would hide under my quilt for hours until I finally fell asleep. I didn't know what the strange images were. My parents didn't either. They thought that perhaps I was having nightmares, or worse; maybe I was hallucinating.

It was not until high school that I learned my condition had a name: synesthesia. I was doing a science project on colors and came across the phenomenon in the literature I was researching. From then on I was fascinated by it, although it would take many years before I began studying the condition in my own lab. And then one day, the synesthesia that had caused so much fear in my earlier years saved my life.

In March 2008 I was in Australia finishing up my postdoc in eminent consciousness at the philosopher David Chalmers's Centre for Consciousness. On this particular day I was hiking in a rain forest. Snakes were sunbathing everywhere—poisonous eastern brown snakes. These snakes are considered the second-most venomous land snakes in the world, with a neurotoxic venom that can cause death to humans within half an hour if the person is left untreated. Even the baby snakes have enough venom to kill an adult human. When hiking in the rain forest there is usually no way of quickly receiving antivenom, so avoiding snakebites is paramount. Usually, avoiding the snakes is easy. The hungry and alert snakes slither away when they hear footsteps. But the satisfied and lazy sunbathing snakes don't feel like moving at all. Occasionally you will need to make them move by throwing little sticks at them, because the trail is so narrow that you can't step around them.

As I was hiking down a narrow trail, suddenly I couldn't see anything but a field of bluish-green rotation projected out in front of me. It completely blocked my visual field and I was forced to stop. This had never happened before. I hadn't experienced any fear, yet my visual field was one big landscape of synesthetic fear.

After a few seconds the colors faded enough for me to see through them. And there it was: a curled-up brown snake less than two feet from where I stood. It was huge in comparison to most of the brown snakes in the rain forest, and it was hissing at me. I felt the panic in my chest.

"Don't move," I told myself.

The snake kept hissing.

I wanted to run. But I knew that running was the wrong thing to do. If you run, you may scare the snake and it may attack, and if you get bitten and you are running, the venom will kill you faster.

"Don't move," I kept telling myself.

Somehow I managed to stand completely still until the snake slithered off into the trees. Maybe it was only ten minutes, but it felt like hours.

That was when I started loving my fear synesthesia. It had saved my life in a moment when my conscious awareness hadn't even perceived a threat.

The information that normally remains behind closed doors isn't limited to potentially dangerous stimuli. As we will see later in this book, synesthesia can provide individuals with seemingly amazing skills in almost every domain. There are synesthetic sculptors, mathematicians, pianists, and even self-proclaimed psychics. Years later I'd discover that this is but one example of how the brain is constantly processing much more information than what we are conscious of. In the case of synesthesia, some individuals have found a way to tap into this information stream.

Around the same time as my encounter with the snake, I discovered that my daughter, Rebecca, is a grapheme-to-color synesthete. Letters and numbers have their own unique colors, even when printed in black. One is blue, two is yellow, three is green, four is purple, five is red, and so on. It shouldn't have surprised me that she has synesthesia, as synesthesia runs in families.

I found out that she had the condition when we were taking our regular route to her day care. The short journey included driving under three consecutive highway bridges. My daughter was only three at the time, so each day I would count out loud when we were driving under them. "One bridge . . . two bridges . . . three bridges." She would count aloud as well. She has been obsessed with numbers for as long as I remember, so this was always a very exciting part of our short drive for her. One day after passing the third bridge I just spontaneously asked, "What's the color of three?"

"Green," my daughter replied.

The response came so naturally to her that I immediately doubted that this was an arbitrary answer. When we got home that night I asked her about the numbers from zero to ten, and all the letters in the English alphabet, as well as the three additional letters of the Danish alphabet (my daughter is bilingual). She had unique colors associated with all the graphemes, except zero. For example, *J* is magenta. Zero became colored a few years later and has changed its color since then (it started out as golden and is now blue). She is eleven now and the other numbers and letters have retained their original colors or slight variations on them.

Many years later it occurred to me that perhaps my parents were synesthetes, too. One would think I'd have asked them earlier, but I never did. I finally asked during a visit to Denmark. My mom was cooking. I casually walked up to her and asked, "What color is seven?"

She looked at me as if I was from a different planet.

"Are you all right?" she asked, concerned.

I nodded. "Yeah, I'm great."

"Let me know if I need to make you some chicken soup," she said.

When I heard my dad enter from the yard, I hurried into the washroom to meet him. "What color is seven?" I asked.

He too looked at me as if I was from another planet, then shook his head slightly and said, "White, of course. Why?"

"What about *G*?"

"Brown. Why? Why are you asking this?"

At a mature age my dad was very surprised when I told him that not everyone sees numbers and letters as colored. He has never read any of my work on synesthesia. I write in English, and his native language is Danish. And I don't think we had ever talked about my condition since I was a young child when no one knew what was going on. In our family most dinner conversations are about politics, psychopaths, foreign countries, vacations, art, theater, fine wine, and grandchildren. I told my dad that I had a different form of the condition, and though he vaguely recalled me talking about it as a child, my form of it is so different from his that it never did ring a bell.

After my experiences in Australia I was even more fascinated with synesthesia. Not only was it aesthetically gratifying, it had quite literally saved my life, and it was abundantly apparent that it powered my daughter's prodigious memory. I wondered what it could teach us about the human brain, and what role synesthesia might play in accessing hidden cognitive abilities in all of us. So after finishing my postdoc in Australia I returned to the States and started a lab devoted to the study of synesthesia and special talent.

Synesthesia as a Tool for Enhanced Mental Abilities

Synesthesia appears to enable people to enhance their memory capacities and to perform mental feats related to their synesthetic abilities.

Our synesthete Megan remembers birthdays, meetings, activities, and so on, by consulting the virtual calendar that circles her body. The rest of us pull out our paper or online calendars to jot down plans for the future. But as a time-space synesthete, Megan can use her mental calendar as a tool for memory. She looks straight ahead and can tell you precisely what days work for her.

I recently discovered that I remember information on websites because sites that I frequent or that impress me get located in virtual space circling my body. My current university's e-mail login page is just after ten if you envisage my personal space as a clock. I go to the virtual space to retrieve material I have read, and this can also help me relocate the site in "real" virtual space if I need to double-check.

But this isn't an ability unique to synesthetes. Anyone can learn to associate qualities such as colors, personalities, or locations with information that they need to access quickly. For instance, you might "color code" phone numbers, lists of data on which you're being tested, and verb declensions and key phrases in languages you're learning. We'll look closer in chapter 4 at how to go about acquiring synesthesia-like associations and developing a superhuman memory.

In addition to aiding memory, research shows that synesthesia can enhance other abilities as well. On top of the "symphonic massage" that Megan feels when she hears music, her peculiar tactile sense has enabled her to feel other people's pain—pain that we normally would only see in the form of a facial expression. She regularly relies on this ability in her work as a nurse.

Some synesthetes apparently have superhuman insights into the thoughts and feelings of their fellow humans, others are equipped with incredible memory skills or admirable artistic acumen, and yet others are able to influence stubborn minds by instantiating their atypical sensory connections in work environments or commercial settings. Research also shows that synesthesia turns up especially often among artists and geniuses, and that the condition plays a crucial role in facilitating their extraordinary acumen.

So what's the connection between synesthesia and supermind ability? Our lab's hypothesis is that synesthesia can be a basic building block of potential supermind ability. Let's look at how this may be the case.

All brains already perform extremely complex calculations when we complete the simplest of tasks. A two-year-old can proficiently maneuver a computer mouse with one hand while being fully focused on what is presented on the screen. The underlying calculations the brain needs to perform to make this happen are so complex that few of us would know how to consciously perform them, even with access to a supercomputer. To move a computer mouse, the brain must quickly calculate and recalculate distances that the mouse needs to move for the cursor to move different distances. We are never consciously aware of these complex calculations and hence cannot use them to solve mathematical problems at will. In some people, however, *synesthetic images often function as a translational device between unconscious calculations and the solutions to difficult problems.* For example, British savant and synesthete Daniel Tammet sees the product of large digits as the shape fitting in between the images of the digits. This allows him to multiply faster than a calculator.

Synesthesia, however, is not reserved for the lucky few who were born with it. It is an extreme variant of multisensory integration—the way the brain combines the information coming in through the five senses to create a single conscious experience. Sensory information first enters through channels specialized for a particular sensory modality, such as hearing, taste, or vision. Even though each channel is more or less independent, the brain attempts to match up the timing of each signal so that the experience makes sense, making it impossible for you to experience any particular sense in isolation. Sometimes the information streams coming from two channels don't quite match up. One superb example of multisensory integration is the McGurk effect, a phenomenon in which auditory experience can be manipulated by a change in what the listener sees. When integrating visual and auditory signals, the brain tries to make sense of conflicting information, sometimes allowing one of the signals to win out. For example, if you watch a video clip of a man repeatedly saying "fa, fa, fa" dubbed over with an audio recording of "bah, bah, bah," the visual experience will win out, causing you to actually hear "fa, fa, fa."

Because synesthesia is a quirk of multisensory processing, it is possible for all of us to obtain the intellectual and creative advantages of the condition by reinforcing naturally occurring integrative processes. The reasoning behind this is simple. Synesthesia is the result of new, unusual brain connections, such as the connection between sounds and colors, or sounds and emotions. And it turns out that it is quite possible to *learn* to become synesthetic to some extent, by training the brain to associate, say, colors with memories or emotions, or colors with sounds and graphemes. While synesthetes typically are born with these unusual linkages, nonsynesthetes can actually rewire their brains through deliberate practice, building the same sorts of unusual bridges.

It is well known that memorizing arbitrary facts, unlike memorizing synesthetic connections or other meaningful associations, can be extremely difficult. Part of the reason for this is that memories are stored in fragments in different regions of the brain. When you retrieve a memory, the hippocampus—the brain's main memory center—assists in putting together the fragments, sometimes correctly, sometimes incorrectly. In the case of emotional memories, the neural networks connecting the memory fragments are tied to the amygdala, the brain's fear center, and to other neural regions involved in processing emotions. Memories tied to emotions tend to be easily retrievable and very intense. This is why it is so difficult to recover from post-traumatic stress disorder or to get over the breakup of a relationship tied to a lot of memories. The amygdala intensifies those memories and makes them easier to recall (which is sometimes unwanted, as in the case of a breakup). This tie was an enormous advantage for our ancestors. Not remembering the name of one of the individuals in their tribe wouldn't kill them, but forgetting that tigers are dangerous would. So, the amygdala developed in such a way as to allow for the association of fear with a particularly dangerous animal, thus allowing for quick warnings of impending danger.

Although color associations may be slightly less intense than emotional associations in some individuals, colors play a role somewhat similar to that of emotions. Colors have served our ancestors in numerous ways, and they still serve us today. For example, we react quite intensely to slight color differences in people's faces. A slight variation in the rosiness of a woman's cheeks can make a significant difference as to whether a male study participant will rate her as "somewhat attractive" or "very attractive." The evolutionary reason for this is that rosy cheeks signal not only health but also fertility. Because of these types of phenomena, the brain is very tuned-in to associations that involve color. So, by connecting color with the thing that you need to remember, you can greatly facilitate memory retrieval.

As we will see in chapter 9, the same goes for things like perfect pitch. Perfect pitch is the ability to tell, upon hearing a musical note, whether it is, say, an E or a D. Perhaps that sounds easy. But as a matter of fact, even trained musicians usually cannot do this. Rather than being an innate talent, perfect pitch is at least partially a result of upbringing. A significantly greater number of people who speak tone languages—those in which variations in pitch can change the meanings of words (e.g., Mandarin)—have perfect pitch, compared to Americans (who speak English, which is not a tone language).

Although people don't usually acquire perfect pitch by practicing music, it is possible to acquire it by learning how to associate each note on a piano with a color and an emotion. This basically amounts to learning to be synesthetic (at least in a minimal sense). The science behind this is straightforward. While musical notes are very difficult to remember, colors and emotions are easy to retrieve. So, once new brain connections between sounds and colors, or between sounds and emotions, have been generated, musical notes will be significantly easier to identify and name.

Similar possibilities also apply to other creative activities, such as painting, writing a poem, or building an abstract sculpture. Most poetry relies on a literary technique that is also called "synesthesia." This term should not be confused with synesthesia as a form of perception, but they are superficially related. If a poem states that someone is "hearing blue," for example, that is synesthesia (in the artistic sense). In poetry this may be interpreted as symbolic. By practicing certain forms of associations between, for example, color and sound, people can dramatically increase their abilities to write poetry and paint in interesting and novel ways. No one will become a Yeats or an Eliot without many hours of rigorous practice, but training your brain to become synesthetic can jump-start creative talents that lie dormant.

Why is this so? Because the new associations between, say, color and sound generate new brain connections. The new brain connections seem to allow people to be more creative. This makes good sense, as creativity normally is associated with the ability to combine old information in new ways. And there is indeed a higher incidence of synesthesia among people with creative talent. In fact, synesthesia provides some people with enhanced memory and other cognitive capacities. As researcher Julia Simner explained to the BBC, people with synesthesia, on average, remember double the number of facts from any given period in their life than regular people, who would recall just 39 facts. (One of the synesthetes in the test group recalled 123.) An average person may remember that they went on a vacation to the Virgin Islands when they were nine, whereas the synesthete would remember the name of the hotel, the German kid she played with, and the nice store owners who gave her and her little brother some hard candy. Synesthesia thus tends to provide a general enhancement in memory. It also can facilitate reading speed and mathematical abilities.

A more extreme case of exceptional memory linked to synesthesia is that of Solomon Shereshevsky. He was working for a Moscow newspaper when his editor noticed that he never took notes of addresses, quotes or stories. His synesthesia strongly influenced his memory. It seemed to have no limit in capacity or chronological reach. He could recall long, meaningless lists of nonsense syllables, such as "ma, sa, na, va, na, sa, na . . ." and nonsense mathematical equations after eight years. He was able to remember matrices of fifty digits after only a few minutes of looking at them and could recall them sixteen years later.

Accidental Superhumans

Synesthesia is merely the tip of the iceberg when it comes to super-mind ability. In our research Kristian and I have also studied people who became incredible cognitive jewels by accident. Given the typical stories of people with amazing minds—those of phenomenal artists like Michelangelo and Beethoven, tech moguls like Steve Jobs and Bill Gates, and brilliant mathematicians like Einstein and Gödel—it might come as a surprise that people can turn from completely ordinary individuals into superhumans in a few minutes. But as a matter of fact they can.

Such is the case of Orlando Serrell. He was a perfectly ordinary kid, leading a perfectly ordinary existence for the first ten years of his life. Then in January 1979 a ball smashed into the left side of his head while he was playing baseball with friends. He fell, but eventually got back up and kept playing.. For weeks he suffered from a headache, but he didn't see a doctor and didn't mention his injury to his parents. Eventually the headache ceased. Shortly thereafter Orlando realized that he had the ability to perform calendar calculations about days following his accident. For example, if you ask him about January 15, 2004, he will immediately say "Thursday." He cannot do it for any days prior to the incident. Ask him about any particular day since his 1979 accident and he will likely be able to tell you what the weather was, where he was, and what he was doing.

This condition, which the accident sparked in Orlando, is known as savant syndrome, a phenomenon in which people show extraordinary abilities in a narrow range of capabilities, typically in memory, music, art, calendar calculation, mathematics, or spatial skills.

Researchers at Columbia University conducted an fMRI—a type of scan that shows activity in the brain—of Orlando to test which areas were involved in his newfound skills. They gave him written and verbal dates and asked him to think the answer to the questions. They found that several regions on the left and right side of the brain were involved in the calendar calculations. Apparently the memory centers weren't involved at all, indicating that he actually performed calculations to provide his answers. When Orlando did engage in a memory task by recalling the weather or events that happened on a given day since the accident, researchers found that the areas normally associated with memory were active in Orlando's brain, but that the activity was significantly increased compared to ordinary participants.

There are many other cases of people who become savants by accident. At the tender age of three Alonzo Clemons suffered a severe brain injury after falling on his head. That accident changed his life. It left him mentally and physically disabled with an IQ in the 40–50 range. He couldn't feed or dress himself, and now in his fifties, he still can't. He can't read, write, or drive a car. But the tragic accident did something else to Alonzo's brain. After the incident Alonzo started showing a strong interest in Play-Doh and clay. He was always trying to sculpt things. Even when he did not have access to modeling clay, he was so determined to sculpt that he would use materials he found outside—twigs, grass, leaves.

His mother quickly realized that her child was no ordinary sculptor. After seeing only a fleeting image of an animal on a television screen, Alonzo could sculpt a realistic and anatomically accurate three-dimensional replica of it, correct in each and every detail right down to the muscle fibers.

For more than twenty years Alonzo performed his art in silence and without anyone other than his close family knowing about it. Then in the late 1980s Barry Levinson's film *Rain Man* was released. In the movie, narcissistic yuppie Charlie Babbitt (played by Tom Cruise) learns that he has an autistic savant brother, Raymond (played by Dustin Hoffman), who has extreme memory abilities. Once the movie brought international media attention to savant syndrome, Alonzo's artistic talent was featured in several national media outlets. His sculptures sell like hotcakes, some for tens of thousands of dollars.

Though the syndrome is relatively rare, there are other cases like Alonzo's. For much of his life Tommy McHugh was a petty criminal and a heroin addict. Later he managed to clean up his act and become a builder. When he was fifty-one, he was using the bathroom as he was getting ready for work. He was interrupted by a knock on the door, which caused him to try too hard to move his bowels. The strain led to a sudden increase in blood pressure, and he felt a sudden, sharp pain inside his head. Blood began running from his nose, eyes, and ears, and he collapsed onto the floor. It turned out that he'd suffered strokes in the temporal lobes on both sides of the head and the frontal lobes in the front of the head. It took five hours for surgeons to stop the bleeding but, miraculously, he survived.

After the surgery Tommy complained about dissociative disorder, previously known as "multiple personality disorder." Those with the condition alternate between two or more distinct personalities, usually with impaired recall of the "others." He complained that he felt that he had many people living inside him. "My brain is split in invisible wedges/Leaving many Tommy's on weak crumbling edges," he later wrote in a poem.

When Tommy returned home about two weeks after surgery, he found himself having a sudden urge to write. Before the stroke he had had no interest in the arts. Now, he had an irrepressible urge to be creative. But he had to relearn many skills, including writing. Luckily, his brain injury had not caused any dementia, nor had it impaired his ability to speak. After he relearned to write he started filling notebooks with poetry. Then he realized that he had an urge to paint. In the following months, he drew hundreds of sketches, mainly of asymmetric faces. He then made large-scale paintings on the walls of his house, sometimes covering whole rooms. He could not stop painting. He was obsessed. Eighteen hours of painting a day became the norm. Neurologist Alice Flaherty at Harvard Medical School and imaging specialist Mark Lythgoe at University College London, who did a psychiatric evaluation of Tommy, found that the stroke had given rise to cognitive and verbal disinhibition. When Tommy sees or hears something, it triggers a flood of associations that he cannot easily control. He uses his art partially as an outlet for these associations.

One of our most recent subjects, Tom Jacobson, also had his brain seriously beaten. Tom graduated high school in 1975 with a 1.25 GPA. School was the last thing on his mind. In 1977 he tested out a prototype hang glider that caused him to crash, leaving him with a severe concussion and a badly broken jaw. About a week after he left the hospital, as he was starting to recover, he announced to his parents that he wanted to go to college. His parents were baffled, as college hitherto had been the least likely path for Tom. But Tom had changed. He moved through junior college in three semesters, completing calculus with an A, and entered law school at the age of twenty-one. After reading about complex systems theory, a field that seeks to unify the sciences with mathematical theory, Tom began seeing connections between chaotic and complex systems and cosmology and is now recognized for his contributions to astrophysics, despite never having obtained a degree in physics.

Patrick Fagerberg acquired savant abilities while partaking in less dramatic activities. He was attending a concert in Austin, Texas, when a thirty-foot steel camera boom fell, hitting him directly on the head. His traumatic brain injury ended his successful law career, but during the following year he suddenly developed an urge to paint—and displayed an extraordinary talent for it. Painting became his life, and the world is starting to notice. Gremillion & Co., a major gallery in Houston, Texas, recently accepted him as a client and is showing his art. Patrick says that he thinks about painting all of his waking hours. Not a moment goes by when it is not on his mind.

Another recent subject of ours, Leigh Erceg, fell forty feet down a 75-degree inclined steep slope while feeding chickens, hit her head on numerous rocks, and sustained gashes from those impacts. She was flown to a trauma hospital. She had a cervical spinal injury, as well as fractured teeth. She was initially paralyzed from the neck down. Remarkably, she recovered from the spinal injuries in a few weeks, following surgery. Subsequently, she had facial reconstruction and numerous tooth implants. After the incident she began to see the world differently. She had no desire to look after farm animals, or any other animal for that matter. Instead she took up painting. She also writes poetry, quickly picks up foreign languages, and has acquired synesthesia.

The cases of Orlando, Alonzo, Tommy, Tom, Patrick, and Leigh show that it's not only special brains that can achieve the ultimate level of cognitive performance. All six went from being ordinary individuals with average intelligence to savants in their special area of talent. What's interesting about such cases is that they seem to be not far away at all from how the normal or "neurotypical" brain functions. In other words, superhuman ability is not an otherworldly and impenetrable attribute of arcane brains. It's not a God-given knack, but something closer to an innate aptitude that lies dormant in the great majority of us. Acquired savants show us that ordinary brains have hidden superhuman abilities. We just can't usually access these abilities. Brain injury, in certain cases, facilitates access to otherwise inaccessible pockets of talent.

And what does this mean for the rest of us, who don't want to meet with injury or stroke? It's the means by which acquired savants gain access to those hidden talents and abilities that are so instructive. By studying cases like those of Orlando, Alonzo, Tommy, Tom, Patrick, and Leigh, we learn how the brain must be changed in order to tap into that hidden reserve, and we can take that information and develop systems, drugs, and technologies that help us achieve the same results—without the risks associated with injury and the myriad complications that come from disease.

Disabling the Brain Boss

Sometimes extraordinary ability is due to a head injury as in the cases of Serrell and Clemons; sometimes it's due to a stroke; and sometimes it's due to an even more unexpected source. In 1986, a Canadian biologist named Anne Adams took a leave of absence from her job after her son got into a bad car accident (from which he would eventually recover). With more time on her hands, she started painting and found herself not only good at it, but absolutely fascinated by it. Soon Anne found herself spending every day from nine to five in the art studio. It didn't take long before she quit biology altogether and became a full-time painter.

Then in 1994 Anne became obsessed with the music of the composer Maurice Ravel. She was particularly entranced by Ravel's famous one-movement orchestral piece *Boléro*. It's one of the best known compositions in classical music. It alternates between two main melodic themes, repeating the pair eight times over 340 bars with increasing volume and layers of instruments. She decided to paint an elaborate visual translation of the composition on canvas, calling it *Unraveling Boléro*.

What Anne didn't know at the time was that her newfound talent for painting was due to the frontal areas of her brain slowly unraveling. Frontotemporal dementia, which differs from the memory-related form of dementia known as Alzheimer's, often leads to sudden remarkable enhancements in people's artistic abilities. In frontotemporal dementia, cells in the brain's frontal and temporal lobes slowly begin to die off, changing the configuration of certain brain circuits and altering the connections between the front and back parts of the left hemisphere of the brain, as well as the connections between the two hemispheres.

As it turns out Ravel probably was also suffering from frontotemporal dementia when he wrote *Boléro*. After its premiere at the Opéra de Paris in 1928 a woman in the audience was heard shouting that Ravel was mad, to which Ravel responded, "You, my friend, are the only one here who understood the piece." He later told Spanish-Cuban pianist and composer Joaquín Nin that the work had "no form, properly speaking, no development, no or almost no modulation." Ravel was indeed going mad but managed to produce his most celebrated piece in the process.

As we will see in later chapters, what is likely going on in these cases is that the controlling, bossy left hemisphere of the brain gradually loses its ability to regulate the right hemisphere. The left frontal areas of the brain and the frontal regions of the temporal cortex normally keep the creative right brain in check through organized thought and decision-making, thereby suppressing its creative activities. It's micromanagement supreme. When the frontal areas deteriorate, however, the artsy right brain regions are no longer suppressed and become unusually active, very often leading to amazing artistic abilities, especially in the areas of painting and drawing. This is also known in the medical literature as "paradoxical functional facilitation" of the rear right-side parietal cortex, which sits on top of the head, and of the rear part of the right temporal lobe on the side of the head. These regions are used in accurate copying of images and in drawing internally visualized images.

In effect, when the bossy left hemisphere is "shushed" and the creative right brain is allowed to "speak," artistic talent proliferates. In chapter 5, we'll show you ways you can begin to silence that bossy left hemisphere and access areas of the right brain that often remain silent.

Ordinary Brainiacs—Without the Trauma

Given what we've seen, we may all have the potential to become math whizzes or brilliant artists. But we can't very well go around attempting to bang our heads in just the right way to get these amazing abilities. Nor can we wait for a stroke or frontotemporal dementia to set in. Even if we could occasion these traumas, brain injury, strokes, and dementia come with so many costs that it's not a trade-off we would be willing to make.

So maybe we cannot become acquired savants and maybe we wouldn't really want to. But we *can* imitate them.

We can use the lessons that these unusual cases have taught us about the brain and apply them to ourselves. Throughout the book we will cover, step by step, a number of strategies for acquiring some of the abilities of a savant. Here's a little taste.

Suppose in 2010 your brand new acquaintance tells you he is turning 40 on Saturday. Within seconds of meeting your new acquaintance, you tell him that he was born on a Friday. How could you accomplish this? By using an algorithm. You quickly divide 40 by 7, get 5 and a remainder of 5. Knowing that 2002 was a leap year you swiftly determine that he has gone through 9 leap years in his life. You subtract 7 from that number and get 2. The algorithm now is to add the remainders of 5 and 2, then add 1 to that and count backwards from Saturday. That yields a Friday. As we will see in chapter 5, by memorizing a few dates and matching weekdays for each month of the current year, you can use the same method to speed-calculate the weekday of any date within this or the last century.

Maybe calendar calculation isn't your thing. Maybe you are struggling to remember everyday things. You only have to buy milk, wine, and oranges and yet when you get to the store, you have forgotten what you needed to buy. Want to improve your memory? The keys to memory are emotions, narrative, and mental imagery. Part of the reason for this is that memories that cause strong emotions and memories of narratives are laid down and stored in a more intense form than regular memory. So, if you are concerned about your inability to remember a simple grocery list containing the items milk, a bottle of wine, and oranges, create an emotional narrative in your mind with different feelings and associations. You might visualize a waterfall of milk in your kitchen. This may give rise to a feeling of absurdity or pleasure. In the living room you can imagine that the bottle of wine cracks on the floor. This will trigger a feeling of instability. As you are walking through the aisles of the store, you start your mental journey in the kitchen. There you are hit with a feeling of absurdity and see a waterfall of milk. As you walk into the living room you suddenly have this uncomfortable feeling of instability. You see the broken glass on the floor and remember to pick up wine. You continue this way until you are done.

Learning algorithms and strategies like the two just presented does not simply help you perform party tricks and impress your new acquaintances. The brain already operates via algorithms and smart strategies. It has evolved to conserve its energy and speed up its computations through a series of shortcuts. Think about an outfielder during a game of baseball. It has long been a mystery how he can actually catch the ball. This certainly isn't possible on the basis of conscious decision-making, for it's too slow. It turns out that the brain doesn't even bother calculating any real facts about the speed or trajectory of the ball. Instead, it uses an algorithm that adjusts the outfielder's running speed so that the ball appears to continually move in a straight line in his field of vision. In other words, through practice, the outfielder's brain has developed its own formula to make it possible for him to catch the ball. It's a shortcut.

Why can't we all do it? It's because not all of us have internalized that algorithm so as to make it automatic. But you can purposely internalize algorithms through targeted training. You simply have to supplement the brain's normal ways of functioning. The brain quickly "makes room" for the extra algorithms by changing its neural structure. This is also known as brain plasticity.

The Supermind in All of Us

Sometimes extraordinary ability is due to a blow to the head as in the case of Orlando Serrell (a method we don't advocate); sometimes a stroke or the loss of a sense brought it about (ditto); other times it's intentionally ingraining a new shortcut or connection in our brain (methods that require nothing more traumatic than dedicated practice). Our conclusion is that these various methods allow us to become conscious of information normally buried within our unconscious minds. We humans—varmints with consciousness—cannot for the most part purposely manipulate information we are not consciously aware of. But by various means it *is* possible to facilitate conscious access to the brain's enormously complex computations, and this is the key to finding and activating extraordinary talent.

In the following chapters, we'll present cases that may initially seem very different from one another. Despite their apparent differences, they all share one common theme: Ingenious or superhuman abilities lie dormant in all of us, and they can be activated and whipped into shape. What we've seen and documented in our own lab is that ordinary people have the brain potential to acquire superminds. Turning the brains of ordinary individuals into extraordinary superbrains may sound like science fiction, but it's not. Science is working on many ways to access and mobilize the potential for prodigious talent within each of us, and the incredible people who have generously given their time to our investigations are helping to show us the way.

CHAPTER 2

Brilliant Impact

*How One Blow to the Head Can Unlock
Hidden Talent in Minutes*

In the right conditions, a baseball bat can become a weapon of devastating destruction—as well as a key to unleashing extraordinary cognitive ability.

On Friday, September 13, 2002, thirty-two-year-old Jason Padgett was attacked by two men as he was leaving a karaoke bar. He was struck twice on the back of the head and lost consciousness for a few moments. The next thing he remembered was being on his knees, as one of the men struck him again and again on the right side of the head.

Adrenaline pumping through his veins, Jason reached out for one of the men and pulled his leg out from under him, driving his teeth into the man's leg. He grabbed the guy's crotch, squeezing and twisting as hard as he could. The man screamed and pulled away from his grip. At this point the other guy started kicking Jason in the head and stomach. "Give us your jacket, shithead!" he screamed while striking Jason again and again. Everything was a blur, and Jason remembers thinking he was going to die. He couldn't see clearly and didn't know if it was because of the blood dribbling down into his eyes or because something was wrong with his vision. Somehow he managed to get the jacket off, and his assailants grabbed it and ran into the distance.

At the emergency room Jason received an X-ray to check for broken bones and answered questions about what happened. A couple of hours later he was diagnosed with a concussion and sent home to rest for a few days. As is typical in these sorts of assault cases, the police had so little information to go on that finding the perpetrators would prove impossible.

Jason barely made it home before he realized that something unusual was happening. Reality was fragmented. The edges of things were sliced into small pieces. When he moved or watched an object move, patterns would form. The light bouncing off a car's window or its shiny paint would explode into an array of triangles. Every object in his apartment—a chair, a table, the kitchen counter—no longer had a smooth, uninterrupted surface. Outlines of things now consisted of little line segments scattered throughout his surroundings.

The way Jason perceived motion had changed as well. He no longer saw objects as moving in a smooth, continuous manner. Instead he perceived motion as a series of still images. If you've ever been in a room with a flashing strobe light, you have an idea of how Jason now experienced reality. "It's as if someone is pressing the pause button on a video very quickly," Jason said.

In short, his boundary perception had changed dramatically. Objects were no longer whole, but fractured into numerous segments, and motion had been fractured into innumerable still images. He hoped that as he recovered, his vision would return to normal, but to his dismay his jagged perception remained.

When we watch a movie, the images smoothly make their way across the screen. The lifelike nature of high-definition motion reproduction is made possible by the fact that our visual system is primed to process updates in a piecemeal fashion. Modern video screens consist of dense arrays of tiny colored light sources, or pixels, that rapidly change color. Every set of changes in the pixels' colors constitutes a new frame, which actually is a still image. But our brains weave these images together to give the appearance of a continuous stream of fluid motion. So we don't normally experience motion as a series of individual snapshots even though we actually receive the information in that way. After his blow to the head, however, Jason saw every individual snapshot.

This condition may be a form of motion blindness called akinetopsia. A previous case of akinetopsia was reported back in 1983 by neuropsychologist Josef Zihl and his colleagues at Germany's Max Planck Institute of Psychiatry. They noticed that one of their patients had a rather unusual condition after suffering a stroke. Mary, as we will call her, reported having difficulties pouring a cup of coffee "because the fluid appeared to be frozen, like a glacier." As she continued to pour, the frozen image would eventually be replaced by another frozen image of a cup overflowing with coffee. Definitely not a good way to start off a morning, but it may wake you up all the same.

It's likely that Jason suffers from a similar type of akinetopsia as Mary. The main difference may be that the "picture frames" change much more quickly for Jason. For Mary, when the glacier image of the coffee pouring into a cup is finally replaced by the next image, it's too late—coffee is all over the place. Jason gets updates quickly enough that he can stop pouring in time. However, we can imagine situations in which Jason might have to react even faster than this. Suppose he is driving a car when the circumstances require he quickly hammer down on the brakes. In this scenario, Jason's "picture frames" may not replace each other quickly enough for him to recognize the need to avert catastrophe.

Jason consulted various doctors and received several MRIs, but no one could explain his altered perception—and no one could make it go away. As you might imagine, his new way of seeing the world was incredibly overwhelming, and at times frightening. To try to gain some relief from the fractal images that arose unbidden any time he perceived light, Jason nailed blankets over his window frames to darken his apartment. Increasingly reluctant to leave his circumscribed environment, he did not return to work and could tolerate only very short visits from friends or family. He spent the next three years in virtual isolation, leaving his apartment only if it was absolutely necessary. "That was my introspective time," he said. Eventually, through reading and learning about his own condition, he felt comfortable enough to rejoin the world.

It was within this period of time that Jason also realized that there was a pattern to the fractured slices of reality he was seeing and that, in fact, they were visual representations of complex mathematical formulas. He seemed to know this instinctively, having had no prior training in mathematics.

One day, frustrated at the difficulty of explaining his altered boundary perception to other people, Jason tried to draw what he was seeing. For his first attempt he tried to capture the pattern of line segments that fanned out from light reflected on a car window. The result was a makeshift circle comprised of an array of triangles, something like a pizza with very thinly cut slices. He called the sketch *Pi*. He shared the drawing with his mother.

"You can't draw a perfect circle," Jason explained. "All you can do is fill in more and more areas."

"Hence pi goes to infinity," his mother said.

Jason told us that was the moment when everything clicked. He realized that the complex geometrical patterns he saw were representations of the irrational number pi. We often think of pi as an infinite string of digits starting with 3.141592. But that number is merely a

numeric approximation of what pi really is: the ratio of a circle's cir-
cumference to its diameter.

As Jason points out, the problem of getting the exact value of pi is
similar to a problem the famous mathematician Benoit Mandelbrot
discovered with measuring the length of Britain's coastline. If the
stick used to measure a coastline is too large, you have to exclude significant portions of the land in your measurement because the stick
can't trace all of the detail. So you reduce the size of the stick to follow
the curves more closely and count more carefully.

But it turns out that theoretically the stick's size can be reduced
an infinite number of times until it's so small that you need an infinite
number of sticks to trace the coastline's edge. Thus the true length of
the perimeter cannot be measured, as all you can do is pick an arbitrary length for the stick and approximate it. This is known as the
coastline paradox. You'll run into the same problem trying to measure
the perimeter of a circle.

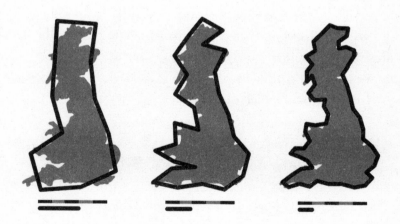

The coastline paradox demonstrates a problem in measuring the
perimeter of curved boundaries. No matter how much you reduce
the unit of measurement, the derived length of a perimeter will
always be an approximation of that perimeter's actual length.

Jason continued drawing circles freehand in an attempt to show others what his new reality looked like. Still none were perfect. Then someone who saw one of his drawings mentioned that it might look quite amazing if only he used a ruler and a compass. The result was awe-inspiring. Jason later proved that his drawing represented pi.

As time went by Jason began to realize that while everyone thought his drawings were fascinating, most people didn't have a clue what he was saying about them. A mathematician told him that he had to learn to speak the language of mathematics if he wanted to make himself understood. So Jason signed up for a trigonometry class and a couple of calculus classes at a local community college. This was a completely new experience for him; prior to the injury he'd been a college dropout whose primary goals had been "to get drunk and get laid." He'd taken a couple of courses in accounting and economics, but the last mathematics course Jason had passed was high school geometry, and he told us he'd copied most of his answers for the final exam. Now he couldn't get enough. He longed to go to school full time but he didn't have the money. However, he continued attending school on a part-time basis and continued drawing complex mathematical figures. Jason eventually began submitting his drawings to competitions and won recognition in 2010 as Best International Newcomer at the Art Basel Miami Beach exhibition.

Synesthesia and Savant Syndrome

The extraordinary changes in Jason's perception following the attack seem to be a form of acquired synesthesia, the kind that has been reported following traumatic brain injury, damage to the brain's white matter (the "wiring" between neurons), stroke, brain tumors, posttraumatic blindness, diseases of the optic nerve and/or chiasm, dementia, seizures, and migraine.

Due to Jason's unique form of synesthesia, he saw complex geometric figures when looking at moving objects. These visual images weren't in his mind, but actually visible, projected out into his surroundings. To get an idea of what it was like when Jason saw these images, consider what happens if you stare into the sun or another bright light—the areas of your retina exposed to the light become oversaturated for a short period of time. The red and orange spots that linger follow your eyes as you move them, so you don't have the feeling that they are part of the actual world outside of your mind, but neither are you just imagining them. You really see them in your visual field. Jason experiences his synesthesia much like these spots, only it has nothing to do with his eyes. It's these synesthetic responses to moving objects that Jason sees and draws so meticulously by hand.

After just a few years of math courses in community college, Jason's unusual synesthetic abilities took on another aspect: He began having similar synesthetic responses to mathematical formulas. Unlike his visual responses to moving objects, his visual responses to mathematical formulas are seen internally or in his "mind's eye." They occur to him the way your vivid visual memories of the person you first fell in love with may occur to you. They are more like images of imagination than true sensory experiences.

Perhaps the most fascinating facet of Jason's metamorphosis is the extraordinary math and drawing talents he acquired. His experience, including his sudden synesthesia, is characteristic of acquired savant syndrome. The leading expert on savant syndrome, psychiatrist Darold A. Treffert, describes the condition as one in which a person has a talent so developed that he or she can perform what may seem like impossible mathematical, linguistic, or artistic tasks. The chief hypothesis is that savant syndrome is caused by a brain lesion or birth defect in the brain's left hemisphere, which results in overcompensation by the right hemisphere. This may allow stereotypical right-hemisphere skills, such as artistic and musical skills and unrestrained or irrational decision-making, to flourish.

Savant syndrome is usually accompanied by severe developmental disorders, most often autism spectrum disorder. In the largest study of savant syndrome to date, 41 out of 51 subjects had been diagnosed with autism. But there are also cases in which savant syndrome occurs without any associated disability, and cases like Jason's in which it is acquired later in life, following central nervous system injury or disease.

A related hypothesis is that we *all* have the skills of savants but that they are not developed because the left hemisphere dominates the average person's brain activity. Typical left-hemisphere skills are thought to include the ability to reason and make rational decisions and to evaluate logical and mathematical problems. So in most people it appears to be a strong adherence to reason and logic that somehow suppresses dormant abilities that we all have in areas of drawing, painting, and music. It's like a stern internal figure that tells us, "You can't make money writing. Become an engineer."

Jason certainly fits the standard characterization of acquired savant syndrome. Prior to the incident that left his brain forever changed, Jason had no special abilities or training in mathematics or art. He scored less than 100 on a standard IQ test, and as 100, by definition, is the exact median score, his results were not particularly impressive. He couldn't even produce a decent enough picture to compete with his peers in a game of Pictionary. They used to tease him about his bad drawing skills.

And despite his lack of prior training, after the assault Jason became obsessed with drawing complex geometrical images using only straight lines. Using his art, he appears to be able to depict compact representations of prime numbers, their "vectors." Some of his visions even coincide with electron interference patterns, a staple of modern physics.

In Jason's case, a brain injury accidentally unleashed the makings of a supermind with enhanced skills in both hemispheres. You could say that sheer dumb luck left Jason Padgett a mathematical and artistic savant.

Love, Luck, and the Ten-Thousand-Hour Path to Success

Acquired savant syndrome appears extremely puzzling in light of some widely received theories of how people succeed in their endeavors. In his book *Outliers*, the eminent journalist Malcolm Gladwell argues that prodigious abilities can be acquired by almost anyone through hard work: ten thousand hours of practice. This is the familiar "ten-thousand-hour rule" first advanced by psychologist Anders Ericsson. That amounts to 417 days. Someone who is obsessed with an activity and practices two-thirds of her waking hours could accomplish prodigious feats in less than two years.

At first glance this may not sound incredibly hard. But few of us would be willing or have the resources to suspend work and family life to practice almost all of our waking hours in order to achieve excellence. It may sound like a dream to become a renowned violinist or a math whiz, but by the second week of having to forsake social encounters and leisurely activities, most of us would give up and return to our old lifestyle. And Gladwell isn't suggesting that we could all put in the ten thousand hours of practice that it would take to gain prodigious skill. It starts somewhere else. It starts with love and luck.

Those who end up practicing ten thousand hours or more tend to start out by being in love with a subject. Bill Gates didn't just pick up programming because he saw a nice future lying ahead of him. He happened to be at a school with a computer and became obsessed with programming. He was so obsessed as a teen that he would sneak out to use a computer available only between two and six a.m. At six a.m. he would sneak back into bed only to be woken up, exhausted, by his mother. Gladwell's idea is that it's this kind of love-like obsession, almost a lovesickness, that drives certain people to practice within their subject area for ten thousand hours or more.

But obsession and practice alone aren't always enough. Uncontrollable factors such as your birth year can have a massive influence on whether you succeed—and we're not talking astrology here. For example, 1920 was a terrible year to be born from the perspective of success. A lot of people were born in the United States in 1920. They grew up during eight years of depression and many were then drafted into the military to serve in World War II. When they finally got out— if they got out—they were middle-aged. On the other hand, 1935 was a terrific year to be born in the United States. There were very few births and hence little competition for schools, colleges, and post-graduation jobs a couple of decades later. Many of the youngsters born in 1935 were taught by elementary teachers with PhDs in hand because of the sharp competition faced by the previous generation.

Such was the case for one of dentistry's greatest pioneers, Garf-ford Broussard. Growing up very poor in the 1930s, Broussard first became interested in dentistry in his early teens when he noticed the most beautiful eyes on a girl with a horrendous set of exceptionally crooked teeth. Thinking she wasn't looking, he slowly raised and lowered his hand, studying how the young girl's appearance was changed so dramatically just by covering her mouth. Startled when the girl caught him playing with her appearance, Broussard vowed to spend his life making sure that no one he met would have to suffer from that sort of deformity.

But the orthodontic care of today was nowhere in sight—the modern brackets, wires, and instruments simply didn't exist. So Broussard created them himself. Enrolled in dental school, he fashioned new tools in the shed behind his home in between study periods. If a new procedure he developed seemed too risky for patients, he'd try it out on himself. His family hardly ever had the opportunity to eat dinner together in the same spot, for the dining table was always covered end-to-end with photographs of patients' teeth. He earned such a reputation as an obsessive inventor that NASA consulted him to explore the potential medical application of a new material they developed, called nitinol. All that hard work paid off—step into any orthodontic office today and you'll see many of Broussard's inventions. Or you might see him. Despite being ninety-three years old, Broussard still operates one of the busiest orthodontic offices in the country, by himself. Call it an obsession. Or a lovesickness.

Broussard himself admits that many of his opportunities emerged from being in the right place at the right time. Would he have become an orthodontist at all had he never noticed the beautiful eyes on that young girl? If he had grown up in the era of modern medicine with fabulous tools at his disposal, would he still have been responsible for so many of the advances in his field? Gladwell's suggestion is that obsession with a subject matter, together with a healthy portion of luck, can impel us to accomplish the ten thousand hours of practice that leads to prodigious abilities, but that it's the time spent practicing that is ultimately responsible for their development.

Obsession and Training as a Route to Mathematical Talent

Gladwell's idea that love, luck, and practice lead to talent has been backed up by scientists and mathematicians. Stanislas Dehaene, a professor at the Collège de France and author of *The Number Sense,* agrees with Gladwell that extraordinary talent ultimately is grounded in obsession and consequent training. This opinion comes from his extensive research of how the brain processes mathematics. He has found that certain basic mathematical skills are not exclusive to humans but are shared with many other mammals and vertebrates. We are all born with a "number sense," the ability to tell how many objects are present in a given location. However, it's not a very exact ability. Nonhuman animals and small children, those who haven't learned to count, can't look at a crowd and say that there are eight people. However, they can distinguish among small quantities. For example, they can detect whether there are one, two, or three things on the table. They can also distinguish between few and many objects. So although they cannot look at a crowd and say that there are exactly eight people, they can tell the difference between eight and twenty-eight people. It's why you have to be so careful handing out that Halloween candy.

The areas of the brain involved in the type of inexact math skill that we are born with are the inferior parietal cortices, located on top of the head. As we learn numerosity, our brain converts number words (*one, two, three, four, five* . . .) and Arabic numerals (1, 2, 3, 4, 5 . . .) into a form that can be stored in its biological wetware. The brain does this by organizing numbers based on their proximity to other numbers. The larger the number is, the fuzzier the representation. So there is a clearer representation of the number 3 in the inferior parietal cortices than there is of 104. This so-called "analogical quantity representation" is a product of evolution. The best survivors were those who could provide good approximations of numerical difference and sameness.

Inexact math is different from what we are taught in school. When we learn to count to one hundred or ten thousand and apply mathematical operations, we engage in a form of exact mathematics. Unlike inexact math, this method is basically a matter of memorizing certain basic facts and algorithms—that is, recipes for manipulating numbers. By memorizing basic facts like specific multiplications ($4 \times 7 = 28$) as well as multiplication and division algorithms, we can methodically multiply and divide large numbers with great precision. But we never acquire a special sense dedicated to this skill.

Dehaene acknowledges that many people with savant syndrome appear to present a counterexample to this theory of how the brain processes numbers. If savants have the ability to calculate numbers accurately with lightning speed, then his theory doesn't seem to apply to all humans. But because savantism is not an isolated phenomenon found in one "freak" individual, the abilities of savants cast doubt on the accuracy of Dehaene's theory.

Dehaene, however, offers the same reply as Gladwell: that the difference in abilities between people with savant syndrome and neurotypical individuals is a function of a difference in training and interest. Because savants are obsessed with the subject matter, Dehaene says, they have more training than most of us. Furthermore, because they think about the subject matter nonstop, they tend to hit upon a few tricks, or shortcuts, as they practice. So, the real difference between them and us is that they are more obsessed with numbers and devote more of their time to studying numbers and math than "normal" individuals do. As far as mathematical savant skills go, Dehaene's suggestion is that some people *appear* to be doing magical calculations because they have discovered a few lesser-known algorithms and spend all day long practicing.

Consider mathematician John Horton Conway, who can tell you the day of the week for any calendar date in under two seconds. How does he do it? He uses an algorithm he developed in 1973 based on what he calls the Doomsday Rule. The rule is based on two particular facts about the Gregorian calendar: It repeats every four hundred years and certain days of a given year always share the same weekday, its "doomsday." After determining a particular anchor day for the century of a given date, one can determine the nearby doomsday for that date's year, which, in turn, can be used to calculate the weekday of the date. Although applying the algorithm seems complicated at first, one can learn it by practicing. Conway does so to this day. In fact, he installed a program that quizzes him on a few dates every time he logs on to his computer. We'll return to this method in chapter 5.

Dehaene thus appears to agree with Gladwell's hypothesis that an obsession with a subject matter and training are the key ingredients in what we often see as brilliant cognitive performance. Even mathematical acumen requires hard work and isn't a reflection of any special innate brain structure. If you love what you are doing, you simply need to learn a few tricks and then practice, practice, practice.

The problem with Gladwell's and Dehaene's views, however, is that they can't explain why *acquired* savants develop extraordinary skills before they have a chance to practice. This was the case for Derek Amato and Tony Cicoria.

"We Didn't Know If God Was in the Room"

One fall day in 2006, Denver native Derek Amato got together with some friends for a pool party. They started playing football in the tiny backyard. Misjudging the depth of the pool, thrill-seeking Derek told his friends to toss him the ball and jumped high over the water to catch it. He crashed headfirst into the hard bottom of the pool's shallow end. "I didn't lose consciousness right away," Derek told us. "I came out of the water and immediately knew I was hurt. I thought my ears were bleeding and I couldn't hear anything. My friends were talking, but I could only see their lips move."

Derek collapsed before his friends could drag him out of the pool. At the hospital he was diagnosed with a concussion and sent home to rest early the next morning. "I think they sent me home just because I was being an asshole," Derek said, laughing. "You know, when you have a head trauma, you get rather frustrated and your triggers are short. I was pretty adamant that I was okay, and I just wanted to go. I actually thought I was in spring training in Arizona for baseball." (Derek *had* been to Arizona for spring training—but years prior to the accident.)

Derek slept almost nonstop the following four days. His mother would wake him up every couple of hours to make sure he was still breathing. While he slept she would put ice on his black eyes and the big bump on his forehead that resulted from the impact with the pool floor. "I looked like I got ran over by a train," Derek said.

On the fifth day he woke up and felt like he was fine. Despite extensive bruising—and despite thinking he was still in Arizona for spring training—he assured his mother he was better and went to his best friend Rick's house.

Derek had never played the piano before, but for some reason he felt drawn to one he knew was in the house. He spontaneously went upstairs and sat down on the wooden bench. His friend didn't know what was going on. Neither did he. "I just felt this weird energy that I wanted to go fumble around with it," Derek said. "I had no idea my hands would know where to go." He put his hands to the piano keys and played like a virtuoso.

"I just sat down and played intensely. It wasn't like someone playing 'Mary Had a Little Lamb'," Derek said with an unexpected insouciance. "It was like Beethoven snuck into my bloodline. All of a sudden someone turned on a switch. I played a classically structured piece. I kept going for six hours." When Derek finally turned around and looked at Rick, his friend had tears in his eyes. "We didn't know what to think," Derek said. "We didn't know if God was in the room. We didn't know if we should have another beer. We didn't know what the hell was going on."

Derek went home that night and went to bed without saying a word about what had happened. He didn't want to *tell* his mom; he wanted to *show* her. The next day, he talked her into going to a music store. She had no idea what he was up to. When they arrived at the store, he sat her down next to him on a piano bench and then played like he had the day before. His mom started crying.

"What are you doing, Derek?" she asked.

"I really don't know, Mom," replied Derek. "I really don't."

A salesperson came over and asked if they wanted to buy the piano. Derek shook his head while continuing to tap the keys. "I'm just starting out," he explained. The clerk left in a huff, assuming Derek was making fun of him.

After Derek told some friends about his sudden ability to play the piano, one of them got ahold of a few people in the Los Angeles music business. They flew him out and put him on stage. The audience was blown away. The organizers were dumbfounded. The guy with zero previous experience really could play. He wasn't making it up. This was Derek's first flirtation with stardom, and he enjoyed the attention. On his return to his hometown in Denver, however, he withdrew from the world, much like Jason Padgett had after his injury and sudden new gifts. "I was trying to understand what my mind was doing," Derek said. "I was trying to get a grip on these changes that really kind of designed a whole new person."

As with Jason, the pool incident left Derek with an acquired form of savant syndrome as well as synesthesia. After the misadventure he started seeing black and white squares going from left to right, a continuous stream of musical notation. "It's like a ticker tape rolling around my brain," Derek explained. It's the constant stream of black and white squares that drives Derek's compulsory finger movements and his urge to get relief through playing. His hands read each square one at a time. Each square represents a musical note corresponding to a finger position on the piano. When he plays he uses six fingers—the thumb, index finger, and ring finger on each hand. "I don't know why all ten don't want to play along," he said. "I suppose it's a comfortable position for my hands."

Derek sees the musical notes in his mind's eye. Though he sees them as black and white, they don't block his outer visual field. They are transparent, like a smudgy window. They can be distracting, constantly racing through his mind. He can't shut them off by closing his eyes or focusing on something else. There is only one thing that can suspend his urge to play altogether and make the notes switch off: the frequent exasperating headaches Derek experiences. Only something *that* excruciatingly painful can give him a little peace of mind. The irony.

There have been other cases of people who acquired musical talent following brain injury. Back in 1994, orthopedic surgeon Tony Cicoria had just finished a phone call with his mother from a public phone booth when lightning struck. In case you've never seen a public phone booth before, the tall boxes and the cumbersome phones they contained were made of metal, an excellent electrical conductor. Metal is likely to become stripped of electrons in a storm, pulling the negatively charged lightning toward it with great force. These two characteristics make for a potentially deadly combination. It's why golfers are hit more frequently than others who are out in the open—the metal in golfing equipment acts as a giant magnet for lightning.

The surge of electricity stopped Tony's heart, soon rendering him unconscious. He also was seriously burned on his face and on his left foot, the entry and exit points for the lightning that shot through his body. By sheer luck, a passerby who worked as a nurse in an intensive care unit saw what happened and resuscitated him.

Tony initially suffered from memory problems after the incident, but they eventually went away. However, at that point he suddenly found himself with an incredible urge to play the piano, something he had never had an interest in doing before. The doctor bought a piano and taught himself how to play. He was hearing music in his head and was composing what he heard. His first composition was called, unsurprisingly, *The Lightning Sonata*. After a few months he added "musician" to his title, traveling around the world to share his newfound talent in between breaks for his clinical work.

None of these acquired savants needed any time to practice algorithms before developing their extraordinary skills. Gladwell and Dehaene both think savants come to master apparently impossible tasks by practicing algorithms. While this may be true for some savants, it is not true for Jason, Derek, or Tony. Their lives were changed almost immediately following their accidents, which suggests that it was the rearrangement of the pieces in their brains that triggered their special abilities.

Looking Inside a Supermind

But if the abilities of acquired savants do not arise from an original obsession with math, drawing, or music and ten thousand hours of training, how do they become so talented? Clearly something must happen in their brains that taps into a potential for talent that was already there. But is that same talent locked inside all of us?

Nearly ten years after Jason was attacked, a local television news station found out about his amazing talent and unusual visions and ran a brief story on him. New York–based journalist Maureen Seaberg, a synesthete herself, found out about his case. Maureen referred Jason to our lab and he soon contacted us to see if we could come up with an explanation for what was going on.

After some initial testing, we decided to conduct a brain imaging study in collaboration with neuroscientists Simo Vanni and Juha Silvanto from the Research Unit and Magnetic Imaging Centre at Aalto University in Finland. We hypothesized that functional magnetic resonance imaging (fMRI) might be the key to understanding Jason's brain. The main difference between a regular MRI study and an fMRI study is that whereas the former merely takes a series of structural images of the brain, the latter shows where the activity is.

Normally fMRI is done by exposing participants to particular stimuli or having them perform simple tasks from inside the scanner, while it records the location of brain activity that occurs during each discrete event. But the nature of Jason's case presented us with a significant hurdle. Brain scanners are tight compartments, and people have to lie extremely still for the images to be clear. Given these constraints it was not possible to have Jason draw inside the scanner, which ruled out testing Jason's brain activity while he was drawing his complex images.

Instead, we focused on the synesthetic images Jason sees in response to mathematical formulas. Though Jason's experiences in response to moving objects and image-inducing formulas differed from more common forms of synesthetic experience, we found that they satisfied the standard characterization of synesthesia—they happen involuntarily and automatically and the images he sees are arbitrary. The most common forms of synesthesia can be identified using something called the Synesthesia Battery, available online, that tests for automaticity and arbitrariness among particular synesthetic experiences.

The battery, however, is not suitable for testing for unusual kinds of synesthesia like the one Jason exhibited. We would have to come up with a similar sort of test to see if Jason could reliably report what he saw, a requirement for diagnosing common forms of the condition. We requested that Jason make drawings of the internal or projected images he saw in response to a range of stationary or moving objects and mathematical formulas, on two occasions approximately three months apart. The drawings he made on these separate occasions were almost identical, which satisfied the first requirement for diagnosing synesthesia.

The visual images that Jason gets in response to moving objects and mathematical formulas are automatic: that is, he does not voluntarily call them up or generate them. He can prevent them from appearing only by closing his eyes, attending to a different aspect of the visual scene, or performing other sorts of distracting maneuvers. Automaticity of this kind is the second requirement for diagnosing the more common forms of synesthesia.

After confirming the diagnosis, we created a list of mathematical formulas that represent complex geometrical images (e.g., $a^2 + b^2 = c^2$) and another list with "nonsense" formulas that did not give rise to any synesthetic response (e.g., $xy = xy^2$). Once inside the brain scanner, Jason was shown the formulas one at a time in quick succession and in a random order.

What we found surprised us: There was increased activity in the left side of the brain in response to the real formulas. Most people who gain special talents from a brain lesion or a defect in brain function have in fact sustained damage to the left hemisphere, unleashing stereotypical right-hemisphere skills like artistic and musical ability. In Jason's case, this was reversed. One of the main blows to his head occurred on the right side, allowing his left hemisphere to work in a new way—generating complex mathematical images in response to formulas and moving objects. It makes sense that this sort of imagery would emerge from the left hemisphere, as mathematical operations seem to take place mostly on that side of the brain.

Likewise, novel imagery, or imagining an object that doesn't actually exist in the world, has been found to be mostly a left-hemisphere activity. In one recent study, for example, participants were shown words that represented two real objects appended together, such as "television-rock," and then asked to try to imagine the synthesis of these two objects. Others were asked to form a novel mental image from two black-and-white line drawings of real objects like a coffee mug and an iPhone. The participants exhibited increased activation in the frontal and temporal cortices in the left hemisphere, indicating that these areas start firing more intensely when the brain is tasked with generating novel imagery that doesn't exist in the real world. For Jason's brain, mathematical formulas may trigger the same mechanism underlying novel mental imagery.

Our study had another unforeseen result. Previous synesthesia research has reported increased activation in certain parts of the visual cortex. The visual cortex is normally responsible for computing color, texture, shape, and other visually perceptible features. But in spite of Jason's very visual synesthesia, we did not find any increased activation in the visual cortex. Instead, the main activity associated with the image-generating formulas occurred in areas of the parietal cortex, located on top of the head, and an area of the temporal cortex, located on the side of the head. The parietal cortex is associated with numerous functions, including preparing for spontaneous action and everyday mathematical activity such as counting. The temporal cortex is associated with processing complex spatial information, among other things. But it's the visual cortex rather than these areas that one would expect to generate visual experiences, including visual synesthesia.

The location of Jason's synesthetic activity explains, in part, why he experiences visual imagery in response to mathematical formulas rather than in response to other, more commonly reported triggers of synesthesia such as individual letters or numbers. The question remains as to whether Jason's visual experiences in fact correspond with some sort of visual representation of the formulas that trigger them. It certainly is possible that the visual responses are generated arbitrarily, much like with the color of graphemes for grapheme-to-color synesthetes. Nonetheless, it's peculiar that the mere recognition of a valid mathematical formula (but not a nonsense formula) changes the way in which Jason's brain processes this information.

Other recent imaging studies of savant cases have also found visual synesthesia without increased activity in the visual areas of the brain. Neuroscientist Daniel Bor and his colleagues from the Sackler Centre for Consciousness Science at the University of Sussex scanned British savant Daniel Tammet, a mathematical savant who can calculate five-digit numbers faster than you can type them in on a calculator. Tammet also holds the European record in reciting pi from memory, having made it all the way to the 22,514th decimal place in five hours and nine minutes. Tammet's savant skills are facilitated by his version of synesthesia: He sees numbers as having colors, shapes, and textures. For any two incredibly large numbers, he sees two colored, three-dimensional patterns, and the product of the two digits is the shape that fits in between the two patterns. He doesn't experience the computational process itself, but merely translates what he sees back into digits.

The mental processing underlying Tammet's synesthesia has certain aspects in common with what we found in Jason. Despite the fact that Tammet reports intensely colored synesthetic images associated with numbers, the brain scan did not show any significant activity in Tammet's visual cortex in response to his synesthetic experiences. Bor and his colleagues therefore suggest that Tammet may have a type of synesthesia distinct from the perceptual form that is more commonly studied. But the findings lend evidence to an even stronger hypothesis, namely that areas of the brain outside of the visual cortex, particularly regions of the parietal cortex on the top of the head, are involved in producing Tammet's colored synesthetic representations of numbers. This shows that the brain is even more plastic than we have hitherto thought. Even color and texture can apparently be generated by brain regions outside the visual cortex.

Although Tammet has had his synesthesia and savant syndrome since he was a young child, his synesthesia is unlike standard cases. He acquired it as a result of epileptic seizures, which are known to damage the brain, sometimes in ways similar to traumatic brain injury.

So, it is not surprising that there are certain parallels between the ways Tammet's and Jason's brains generate synesthetic experiences. Their capacities to process color and texture in higher brain regions, which normally are devoted to other types of neural processing, are quite similar.

In Jason's case the findings tell us that his acquired synesthesia correlates with increased brain activity in the parietal cortex, the brain region associated with math skills. This enhanced activity appears to underlie not only his synesthetic experiences but also his newly acquired mathematical talent. Jason's synesthesia and his extraordinary savant abilities are connected in several ways: the very same brain regions are involved in processing both mathematical information and visual information in a way that is different, and evidentially more efficient, from the way that this information is processed in ordinary individuals. The synesthetic experiences furthermore allow Jason to think about math in a new way. He visualizes mathematical formulas effortlessly and automatically, and hence does not have to make do with abstract mathematical thinking. In other words, his brain provides him with an automatic, visual representation of mathematical connections. It's a bit like having a graphic math textbook implanted in your brain—one that can visually demonstrate the most abstract of concepts. The only effort required for Jason is to look.

Derek, likewise, merely needs to look at the black and white notes in his mind's eye and follow the musical score his brain presents. While we don't have fMRI images of Derek's brain, he did receive the more common structural MRI scan when he was filmed for an episode of the Discovery Channel's *Ingenious Minds*. The scan, however, did not yield any definitive answers. The Mayo Clinic neurologist who examined Derek for the documentary, Dr. Andrew Reeves, identified a little white spot in the parietal cortex on the top of the head and some very tiny spots in the temporal lobe on the side of the head and in the frontal lobe. Derek's head injury from the pool incident wasn't his first. Being active in sports as a child, especially baseball and karate, he sustained several concussions including one resulting from a fall off the back of a truck. Reeves thinks the white spots on the MRI are tiny bits of scar tissue from those previous concussions. He didn't find any other notable structural damage.

The reason behind the lack of any definitive results may have to do with the sensitivity of the imaging technique. Typical MRI scanners are relatively insensitive devices that only pick up tissue deformation and scarring from congenital deformities, concussions, strokes, brain tumors, or other serious brain-disfiguring incidents. Software can determine the size of different brain regions on the basis of the MRI data, but MRIs do not show how the brain's neurons interact, their firing rate, or whether certain brain regions have diminished or enhanced functionality during certain tasks. Gathering that sort of data involves using specialized equipment that requires very targeted investigation. In other words, it can be especially hard to find what you're looking for if you don't already know what it is.

After tentatively concluding that Derek didn't have any major brain abnormalities but merely a "brain that has been smashed around," Reeves suggested a possible biological basis of Derek's condition—excessive neurotransmitter activity. He proposed that if Derek wanted to end his incessant urge to play music, he could try an antiepileptic

drug. Those types of drugs are most commonly prescribed for seizures. Most of them work by slowing down the excitatory neurotransmitter glutamate or by enhancing an inhibitory neurotransmitter called
gamma-Aminobutyric acid, or GABA. When GABA binds to the receptors of neurons, it causes them to slow down. Slowing down brain
activity, however, could also take away Derek's newfound talent. Naturally, Derek declined the treatment.

So what's going on here? Though we do not yet know the exact
answer, there is at least one plausible explanation. A significant portion of the parietal cortex, the region responsible for generating spoken language, is also involved in generating music. This isn't surprising,
for music and spoken language are both rhythmic and melodic in
structure. Just as the tone and tempo of a musical piece sets the mood,
the same elements of a spoken conversation can mean the difference
between devotion and divorce. Furthermore, both spoken language
and music are built upon a limited number of discrete elements combined in a meaningful way. This suggests that musical ability is an
integral part of the brain. Musical savantism might lie dormant in all
of us, brought to the surface through deliberate practice or, as in Derek's case, activated if the brain undergoes massive restructuring.

Does this mean that the hardware underlying the parietal
cortex—hardware that we all have—is the seat of savantlike abilities?
Or does some other brain region deserve credit? At this point it's too
early to tell. Owing to the rarity of acquired savant cases, we simply
don't have enough data to know which particular brain regions, if any,
are the source of supermind abilities.

There's also much to be learned about musical ability in general.
For reasons unknown, musical ability has been studied far less than
many of the other human mental faculties. But some recent work
provides us with a few interesting discussion points. One study lever-
aged the fact that music often doesn't sound good if its structure is

completely arbitrary. Researchers compared the brains of musicians and nonmusicians when hearing chords that were unexpected given the preceding musical context. They found that the brains of both musicians and nonmusicians reacted to awkward transitions in the same way, suggesting certain musical acumen may manifest without training. The researchers followed up with another study showing that both types of brains are also equally unable to ignore musical "errors" while focusing on other tasks. These results suggest that certain musical abilities are innate.

But the brains of musicians and nonmusicians are not exactly the same. In fact, they differ in several important ways, most apparently in the regions critical for musical performance. Harvard neuroscientists Christian Gaser and Gottfried Schlaug studied the differences in gray and white matter between musicians and nonmusicians. Gray matter is the "meat" of the brain, home to all of the hardworking neurons that give neural networks their oomph. White matter is a fatty tissue that consists of tiny insulating nerve endings and glial cells, which perform various brain functions through chemical rather than electrical transmission. The study found that the higher the professional status of the musician and therefore the more practiced they were, the more gray matter there was in the areas responsible for motor function, sensation, and language processing. Researchers believe these differences are *not* innate, because they involve multiple specific brain regions. Typically, with the right kind of practice these areas undergo structural changes to facilitate the demands of musical performance. In Derek's case, structural changes from an injury in effect gave him a "free pass" on the need to practice.

Functional MRI studies have also shown differences between the nonmusicians' and musicians' brains. German neurologist Timo Krings and colleagues studied the differences in brain activation between professional piano players and controls, finding that the professionals actually use *less* of their brains to initiate the complex movement necessary to perform. One explanation of this is that long-term practice changes the way in which the brain executes complex repetitive movement—the brain is triggered to reorganize so that it can play more efficiently, using minimal resources. Thus, although the brain appears to be structured in a particular way that facilitates some musical talent, this structure may be altered with practice that stimulates the appropriate regions. Derek's case shows us that these changes aren't always necessary to achieve proficient musical ability. You could say that his brain is configured like a practiced musician's, although for Derek it wasn't years of practice that gave rise to his extraordinary talent, but rather one extraordinary event.

A Flood of Talent

It will take many years of research before we have a more complete understanding of what actually goes on in the brains of people who acquire superhuman minds after brain injury. But we have some clues as to how a brain injury might trigger the reorganization of affected regions. The existing scientific evidence suggests that acquired extraordinary talent, much like learned talent, is due to changes in the brain's connectivity. Lesions can affect both the gray matter, which assimilates information into conscious experiences, thoughts, and actions, as well as the fatty white matter, which is involved in high-speed transfer of information between different parts of the brain. Gray matter is more plastic—there the brain can heal minor lesions with more ease. White matter damage is normally considered the most severe type of damage, as the brain is less adept at repairing its own wiring and therefore cannot easily transfer the function of one region of white matter to another.

In cases in which there is no hope of repairing a damaged area, however, the processing normally performed in one region can sometimes be performed in another. And it's when one region takes up the duties of another that new abilities can sometimes be gained. Trauma to the brain's white and gray matter often severs connections that normally facilitate communication among different regions. Cases of acquired savant syndrome like Jason's and Derek's, however, suggest that some brains end up with increased or strengthened connectivity as a result of remodeling that takes place during the healing process. This was confirmed by CUNY neuroscientist Tony Ro and colleagues in a case in which a woman developed synesthesia after a stroke. The stroke affected the neural connections between the auditory cortex (the region of the brain responsible for hearing) and the somatosensory cortex (which is responsible for touch). After the stroke the woman could literally feel sounds in her body. This wasn't simply a case in which she could feel deep sound vibrations of drums and bass when listening to loud music. Sounds made her feel as though something was touching her. The researchers found that connections between the auditory cortex and the somatosensory cortex in the woman's brain were both more numerous and strengthened, allowing the auditory cortex to transmit information directly to the somatosensory cortex, which in turn led to a feeling of touch in response to sounds. In this case it is likely that the somatosensory cortex was deprived of some of its normal inputs due to the stroke, triggering the formation of new connections between the two regions.

Jason's and Derek's cases suggest a hypothesis for explaining extraordinary mental abilities that come about as a result of head injuries. Traumatic brain injury (TBI) occurs when external forces injure the brain, either as a result of blunt force trauma or shock waves from a blast. In both situations, the inside of the accelerated skull comes into contact with one side of the brain, generating a secondary shock wave throughout the soft tissue. If the force is strong enough, it can cause the brain to "bounce" off the other side of the skull, resulting in another shock wave. The waves emanating through the brain twist and pull on the connections between neurons, tearing them apart, causing damage to various areas. Depending on the severity of the shock wave, TBI can be very extensive, and multiple TBI incidents can have compounding effects.

The potential long-term effects of TBI have recently come to the forefront of public discussion after the landmark $765 million settlement between professional football players and the NFL. Repeated blunt force trauma to the head and the ensuing concussions can cause chronic traumatic encephalopathy, or CTE, a degenerative disease that leads to near-total mental debilitation. Symptoms of CTE usually take many years to appear and usually start out subtly as apathy, depression, or irritability. As the disease progresses, its unfortunate victims may experience tremors, speech problems, and dementia. Many career football players have exhibited symptoms of the disease, and it's a particularly widespread problem for soldiers who sustain mortar shell attacks at close and mid range. Many of them report memory problems years later.

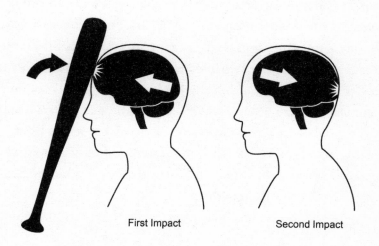

First Impact   Second Impact

Physical force to the head causes the brain to hit the inside of the skull in several places.

The explanation for how a bump to the head might produce exceptional talent has to do with what happens to the brain as it sustains an injury. During a concussion, the nerve function of several distinct brain regions becomes paralyzed as a result of the brain bumping into the skull as it shakes inside the head. Some of the cells eventually die, giving rise to a huge, uncontrolled release of neurotransmitters, particularly serotonin and glutamate. These neurotransmitters bombard neighboring neurons and overstimulate them. It is believed that this overstimulation effect can happen with mild traumatic brain injury, as the release of neurotransmitters is large even with minor lesions. It also happens after other types of brain injury, such as stroke, that cause neurons to die off.

The wave of neurotransmitters released from dying neurons massively enhances brain activity in neighboring brain regions, potentially triggering long-lasting structural changes. This is because the two main neurotransmitters that flood the brain after injury, serotonin and glutamate, both enhance neural activity.

Neurotransmitter flooding may also explain the formation of new ways of pairing information after brain injury, leading to conditions like synesthesia. Visual imagery is by far the most common way for the brain to represent the world, so it's not surprising that if a network of new connections gives rise to conscious experiences, these experiences are visual in nature. After being beaten up, Jason experienced visual images as complex mathematical patterns, and after the impact with the pool floor Derek experienced visual images of black-and-white musical notes. These visual images make it possible for the two unschooled whizzes to partake in artistic activity in ways that would not otherwise be available to them.

A Superhuman Mind Hidden Inside All of Us

Cases of acquired savant syndrome show us that the brain is primed with immense talent. This goes against Gladwell's and Dehaene's hypotheses about the basis for savant syndrome: that the phenomenon is nothing more than obsession with a particular domain followed by intense training. In acquired cases, individuals who sustain mild traumatic brain injury can wake up with the sudden ability to perform difficult tasks normally reserved for the practiced professional. Changes in levels of neurotransmitters yielded access to abilities that were already waiting to unfold. What all this means is that somewhere deep down in our brains, most of us have seemingly superhuman capacities. You could say that we are born equipped with the means to exhibit extraordinary mental abilities.

This raises the question of why we don't all have extraordinary abilities. The answer is that the neurotypical human brain suffers from the limitations of our conscious abilities and by inhibition imposed by dominant brain regions or, in some cases, the whole left hemisphere. Although the potential is there, most of us cannot consciously make huge calculations in our heads or perform calculations that spit out complex mathematical images in our field of vision.

But don't go out and bang yourself up. A blow to the head only rarely causes the right sort of changes that can lead to savant syndrome. Why is that? Why do individuals like Jason and Derek acquire extraordinary skills after brain injury when the overwhelming majority of others suffer only negative consequences?

Two factors play a role here. One is the location of the injury. When lesions occur in regions that process information from the senses or are involved in mental imagery, sensory conditions like synesthesia are more likely to develop. Another factor is how plastic an individual's brain is. All brains are not equally likely to undergo the same changes, even after sustaining similar injuries. Research on brain injury in children has shown that alterations in neurotransmission during the critical period when a brain region undergoes most changes can promote the outgrowth of abnormal neural connections. Since it is likely that there are significant differences in the plasticity of the adult brain as well, some people are bound to be more susceptible to the formation of new neural connections than others, particularly young and healthy individuals but also individuals with preexisting unusual brain connections, such as synesthetes.

However, if we find out how to change the connections between neural regions, then these super-abilities might be available to most people. As we will see throughout this book, the future looks promising in terms of providing technology and medications that can bring about the type of neural rewiring that could unleash the hidden talent waiting to be uncovered in all of us—without the traumatic injuries that Jason and Derek experienced.

CHAPTER 3

A Flexible Mind

Restructuring the Brain

At the age of eleven, Carly Fleischmann typed a simple message on her father's laptop: "Hurt. Help." Normally this sort of message would not be impressive if written by an eleven-year-old. But for Carly it was different—she suffers from severe autism and had never before spoken or written or been able to connect to the world.

Until that day, Carly was thought to be mentally handicapped. She regularly threw temper tantrums, thrashing her arms and slamming them on the table. So the message she typed startled her parents. Until Carly's message, they assumed that Carly couldn't understand what was being said around her and couldn't have responded in any meaningful way. Doctors told them that if she was lucky she might eventually reach the developmental level of a six-year-old. The message Carly typed came as a shock. Could she have heard and comprehended their discussions of her handicap over the years?

The answer turned out to be yes. Carly has autism spectrum disorder (ASD), a developmental disorder normally characterized by regimented behavior, preoccupation with details and deficiencies, and delays in social interaction and communication. Owing to her ASD Carly simply hadn't had the means to show that she was indeed listening, taking in, and understanding what was happening around her. After Carly typed those two words, her parents dedicated all their waking hours to helping her practice communicating through the keyboard. She became an extremely articulate writer, expressing herself in a much more sophisticated way than her peers. Three years later she explained herself to the world. "I am autistic. But that is not who I am. Take time to know me, before you judge me. [. . .] I think people get a lot of their information from so-called experts. But if a horse is sick, you don't ask a fish what's wrong with the horse. You go right to the horse's mouth." Carly's unruly behavior, as she explained when she was fourteen, was an attempt to deal with her inner pain and sensory overload. The rocking, arm flapping, screaming, and lack of eye contact, she said, "is a way for us to drown out all sensory input that overloads us all at once. We create output to block out input. Our brains are wired differently. We take in many conversations, images, and sounds at once . . ."

While this case seems remarkable, it turns out that Carly isn't unique. In fact, a new theory of autism predicts that most autistic individuals are much like Carly—they just can't tell us. According to the "Intense World Theory," which we'll describe below, the symptoms and abilities that go along with autism may in fact be the result of greatly increased activity in the brain, rather than reduced activity as was widely believed in the past. As our understanding of autism grows, so does our understanding of not only how people with autism experience the world, but also how it sometimes gives rise to extraordinary savant abilities.

It is commonly reported that *only* about 10 percent of individuals with autism have savant syndrome. A remarkable number of highly talented people are believed to have or have had autistic spectrum disorder, including Albert Einstein, Sir Isaac Newton, Wolfgang Amadeus

Mozart, Andy Warhol, Abraham Lincoln, Temple Grandin, Michelle Dawson, Tim Burton, Hans Christian Andersen, Daryl Hannah, Courtney Love, Dan Aykroyd, Woody Allen, Bob Dylan, Satoshi Tajiri, and Adam Young.

Although this is perfectly consistent with the 10 percent figure, the tight connection between talent and autism gives us some reason to suspect that savant syndrome may be more common than we know, but is underreported because some individuals cannot "report." One thing is clear: Cases like Carly's keep popping up.

For many years during his childhood, Jerry Boutot didn't talk, and used repetitive and self-stimulatory behaviors to calm his overstimulated mind. Although he was nonverbal he mastered any video game in just hours. Now at the age of twenty-two he is a master of creative cartooning. He has his own cartoon series, *Simon and Adrien*, that he posts on YouTube and on his website, simonandadrien.com.

Benjamin Tarasewicz has high-functioning autism. As his mother Malva Freymuth Tarasewicz, a professional violinist, describes in her book *Benjamin Breaking Barriers: Autism: A Journey of Hope*, she started noticing Benjamin's quirks when he was six months old. He was easily hyperstimulated and tensed his whole body in protest. Although he was able to utter some common words, such as *book* and *violin*, at the age of sixteen months, by the time he was three years old he didn't speak at all. It took a lot of intense training for him to learn to speak again and make eye contact, but he went on to become a very social high school student.

Bobby Smith was diagnosed with moderate to severe autism as a young child. At the age of nine he was still not speaking. Instead of using words he used a primitive form of communication with objects. If he wanted to watch television, he would give his parents the remote control. But he was unable to tell them if he was in pain, whether he was sad or happy, what he wanted for Christmas, or what happened at school. The older he got the more difficult he became. He would have complete meltdowns and temper tantrums on a daily basis. Finally, his mother encountered a behavioral therapist who believed Bobby had the ability to speak. When the therapist first started seeing him, he was kicking, hitting, and screaming. So, she began her sessions by teaching Bobby to remain calm. She then taught him to speak by refusing to acknowledge his response until he used his words. His mother used the same technique at home. Bobby now speaks in complete sentences, and his rages and meltdowns have ceased completely. And once his meltdowns had ceased, the family discovered that Bobby was able to imitate nonhuman sounds to perfection, such as the sound of the turn signal in a car.

Then there is Alex Eveleth. "When I was in fifth grade, my brother Alex started correcting my homework," writes Rose Eveleth in *Scientific American*. "This would not have been weird, except that he was in kindergarten—and autistic." Although Alex had severe difficulties with communication, he was clearly a highly talented child. Fifteen years later he still cannot write, but he can identify mistakes in fifth-grade math and English homework.

The abundance of cases of individuals with autism who are highly talented but who cannot communicate this to the world indicates that very many, if not all, people with autism have some extraordinary mental abilities. For those of us who'd like to develop incredible ability, the brains of autistic people can thus shed light on the particular circumstances that unlock extraordinary abilities.

An Asymmetry in the Brain

One common factor among many autistic individuals is an abnormality in the serotonin system. Although serotonin is best known in connection with mood disorders, this profuse neurotransmitter serves numerous functions in the human brain. Serotonin receptors are among the most abundant receptors in the brain. Some of these receptors inhibit neural activity, whereas others increase it. In the visual cortex, for example, increased serotonin can lead to increased neural activity by a coupling to glutamate, the brain's main excitatory chemical.

There is overwhelming evidence that serotonin plays a crucial role in autism. Almost a third of autistic individuals have a 25- to 70-percent increase in blood levels of serotonin, also known as hyperserotonemia. Increased blood levels have also been found to a similar degree in first-degree nonautistic relatives, such as parents and children. As serotonin cannot normally cross over from the blood to the brain in adults, high blood levels of serotonin are not necessarily a good indicator of high levels of serotonin in the brain. High blood levels of serotonin, however, may correspond to brain levels in young children, as the blood-brain barrier is not fully developed until the age of two. Higher rates of autism have also been found in children exposed to drugs that increase serotonin levels, such as cocaine, during pregnancy.

The high levels of serotonin in young children can have a negative impact on the development of serotonin neurons. As serotonin neurons develop and the extracellular levels of the neurotransmitter increase, growth of these neurons is normally curtailed through a negative feedback mechanism, leading to a loss of serotonin terminals. This decrease in the development of serotonin brain cells has also been found in animal studies administering serotonin-increasing antidepressant drugs during pregnancy.

Several studies have suggested that autism may be a syndrome of one of the brain's two hemispheres, with decreased synthesis of serotonin in the left (or sometimes the right) hemisphere. A positron emission tomography (PET) neuroimaging study published in the journal *Annals of Neurology* found decreased serotonin synthesis in the left cortex and thalamus in five of seven autistic boys studied and in the right frontal cortex and thalamus in the two remaining male subjects.

In another PET imaging study of nine autistic girls conducted by the same authors, it was found that four of the subjects had decreased serotonin synthesis in one hemisphere, whereas the others had defects in both hemispheres. The girls with a serotonin deficiency in the left hemisphere scored higher on nonverbal IQ tests than those with defects in both hemispheres, despite other symptoms of autism being similar in the two groups.

More recent PET studies of serotonin synthesis have confirmed that serotonin is generated in different amounts in the two hemispheres in children with autism. The studies found asymmetries of serotonin synthesis affecting the left or right side of the brain. In one study serotonin synthesis was decreased in the frontal lobe in 90 percent of cases. The researchers found significantly increased language impairment in subjects with decreased serotonin synthesis in the left hemisphere compared to individuals with right-hemisphere abnormalities and those with defects in both hemispheres. One possible explanation of the asymmetry is that the early serotonin depletion in the dominant left hemisphere leads to overcompensation in the right hemisphere, though sometimes it's the other way around.

The different patterns of left- versus right-sidedness in individuals with autism line up with the old distinction between autism and Asperger's syndrome. In the old classification, autism required marked language deficits, whereas Asperger's syndrome did not. Many children diagnosed with Asperger's exhibit the same social difficulties, adhere to a rigid routine, and so on, but perform normally or above normal in left-brain activities like complex language, reasoning, mathematics, and sensory integration.

Further evidence for the left-hemisphere theory comes from studies indicating functional improvement with selective serotonin reuptake inhibitors (SSRIs), such as Prozac and Celexa. SSRIs block serotonin transporters, preventing extracellular serotonin from being transported back into the cell. When children with autism are treated with SSRIs, the symptoms they share with people with major depressive disorder and anxiety disorders drastically improve. Depleting the brain of tryptophan, the main protein component needed to make serotonin, on the other hand, results in a worsening of the mood-related symptoms of autism as well as behaviors such as whirling, flapping, rocking, and pacing. Deficiency of tryptophan can also be a contributing factor to a wide range of mood disorders in normal individuals. There is indeed evidence to suggest that autism is symptomatically and genetically related to mood disorders, such as major depressive disorder and bipolar disorder.

A recent study conducted by researchers at Johns Hopkins Hospital indicates that boys born to women who continue using SSRIs such as Lexapro, Celexa, Prozac, Paxil, and Zoloft during the first trimester of pregnancy are three times as likely to be diagnosed with autism. The theory is that the SSRIs interfere with the development of the serotonin system in the fetus, thereby generating lasting heightened levels of serotonin.

The left-hemisphere theory of autism can explain many of the symptoms of autism, most notably the language deficits and delays in language development, as language is a left-brain skill. But it also explains regimented behavior, unusual attention to or avoidance of sensory input, social impairment, and enhanced low-level perceptual memory processing. Studies show enhanced activation in the amygdala, the main fear-processing center, in autistic subjects when they attend to faces or process other kinds of information from the senses. This suggests that autistic individuals may be processing too much emotionally relevant information, owing to depressed serotonin levels. Increased amygdala activity would explain enhanced fear processing and social

impairments, such as difficulties with eye contact, withdrawal behavior, and a seemingly impaired theory of mind—that is, a lack of
understanding of what others might think or feel. Increased amygdala activity might also explain why autistic individuals often engage
in avoidance behavior with respect to certain types of sensory stimuli
and why repetitive movement and regimented behavior seems to
have a calming and stabilizing effect. Avoidance behavior naturally
decreases fear processing, and repetitive movement and regimented
behavior can naturally increase serotonin release from neurons in the
brain stem.

The left-hemisphere theory explains the fact that savant syndrome
occurs in 10 percent of individuals with autism. This supports the leading hypothesis about savant syndrome to the effect that it is caused by
a lesion or birth defect in the left hemisphere that results in overcompensation by the right hemisphere (or in the left hemisphere when the
defect is in the right hemisphere).

When serotonin is eventually suppressed in one side, that is likely
to lead to an increase of serotonin in other brain regions, perhaps in
the opposite hemisphere. This would explain why the brains of autistic individuals tend to have more connections and more activity in
local regions compared to the general population.

It seems, then, that we have a partial answer to why autism sometimes coincides with incredible abilities. Like people with acquired
savant syndrome, autistic individuals have brains that are more connected and more active than people in the general population. The
increased activity and neural connections are often restricted to one
hemisphere, or local regions.

Autism is thus the result of a brain that is locally wired with more
connections than normal. This makes the autistic brain much more sensitive to perceptual stimuli. Some perceptual experience becomes so
painfully intense that the autistic individual can avoid overstimulation

only by limiting attention to the perceptual environment. This explains the lack of normal social interaction that becomes apparent at a young age. But an overconnected brain benefits from an increased capacity to process information, which often leads to exceptional cognitive ability. In other words, a large number of autistic individuals are incredibly bright—most of them just can't tell us.

Autism sheds light on the underpinnings of cognitive talent: the connections between the individual neurons of the brain limit the way information is processed. If there are too few connections, the information takes so long to get to its destination that the brain never finds a solution. But too many connections can overload the brain. Since superhuman abilities emerge within an ideal range of brain activity, the first step in nurturing cognitive talent involves finding that sweet spot: the way to generate new brain connections that can be used to enhance memory and computational abilities without perceptual overload.

"Cocktail Party" Personality and Musical Virtuoso

On July 2, 2000, lyric soprano and accordionist Gloria Lenhoff performed the extremely difficult piece *Knoxville, Summer of 1915*, a lyric rhapsody, at San Diego's East County Performing Arts Center. This was a challenge proposed to her by conductor David Amos. She conquered the piece so exceptionally that he was blown away. He had never heard anyone perform it so beautifully.

What is fascinating is that Gloria has Williams syndrome and is cognitively impaired, with an IQ of 55—unable to make even the simplest of mathematical calculations, like subtracting three from seven. Williams syndrome is a birth defect that causes large malformations of the brain's left hemisphere owing to the spontaneous deletion of twenty-six to twenty-eight genes on chromosome 7. Although this is a tiny percentage of the thirty thousand genes in the brain, even a small defect like this can have huge effects on the body. People with Williams syndrome are physically characterized by a small frame, low muscle tone, and "elfin" or "pixie-like" facial features, such as a small upturned nose, small, widely spaced teeth, long upper lip, a wide mouth with full lips, a small chin, puffy eyes, and a starburst iris. Medically, the condition is characterized by colic during infancy, feeding problems, cardiovascular disease, kidney problems, elevated blood calcium levels, ADHD, developmental delays, and learning disabilities. Williams syndrome patients struggle with spatial relations, mathematics, abstract reasoning, and sophisticated social cues, and like individuals with autism they often suffer from sensory overload. They usually have an IQ of less than 70. What's amazing, however, is that individuals with the syndrome have absolutely fantastic expressive language skills, excellent long-term memory, amazing musical talent, and very social personalities. They are exceptionally friendly, endearing, and polite. They are masters at socializing, sometimes said to have "cocktail party" personalities because of their tendency toward expansive and expressive speech.

Despite suffering from Williams syndrome, Gloria Lenhoff has a repertoire of more than two thousand pieces and has performed nationally and internationally. She sings in twenty-eight languages and in many different styles, and she has perfect pitch. But she is not capable of reading music. She is highly emotionally sensitive to music, breaking out in tears when she hears a sad song or a big smile when she hears a happy song. In addition to her musical skills she speaks fluently in a half dozen languages. "All I do is listen to a person speak in another language, then I know it," she says. She speaks in a highly cultivated way.

Doctors gave up on Gloria when she was only seven. Hardly anyone knew anything about Williams syndrome, which had only recently been given a name. They told her parents that she was deeply mentally retarded and that the best thing to do would be to send her to a home. They didn't do that. She stayed with her parents most of her life and moved quite frequently owing to her father's job in academia. Her parents got her music lessons when she was eleven, but didn't fully realize how extraordinary her talent was until she was thirteen and sang at a synagogue chapel for her bat mitzvah. After that they found the best voice teachers they could.

Williams syndrome more or less reliably results in what neurologist Oliver Sacks has called "an extraordinary mix of gifts and defects." People with Williams syndrome very often are savants. What is it about their brains that gives them these extraordinary gifts? The answer here is very similar to the reason that many individuals with autism are very talented. People with autism typically have defects in the left side of the brain where the language center is located, and they have hyperconnected local regions in sensory areas, the amygdala, and the right hemisphere. Individuals with Williams syndrome also have hyperconnected local regions in sensory areas and the amygdala, which gives rise to sensory and emotional overload. But unlike autistic individuals, their language center in the left temporal lobe is also hyperconnected, and so are regions in the auditory cortex also located in the temporal lobe very close to the language center. However, they have lower connectivity in regions of the brain associated with reading, such as parts of the frontal cortex and the angular and supramarginal gyrus, with mathematics, which has neural correlates in the parietal cortex, and with geometrical tasks in the right temporal lobe. Furthermore, they also have reduced interaction between the two hemispheres. So, as we have seen in many of the other cases, what facilitates savant skills seems to be local hyperconnectivity accompanied by some hypoconnectivity in the controlling left part of the frontal lobe.

The Plastic Brain

But what can we ordinary mortals do to restructure the brain and utilize its hidden potential? Technology and medications that may be able to help us out may not be so far off. But what can we do right now? The answer is that we can learn to use algorithms. An algorithm is a shortcut you use to solve a cognitive problem faster than you could have using ordinary ways of approaching mental exercises. You are already familiar with some algorithms. Consider flash cards: you read a question or problem on one side of the card and try to recall the correct answer or solution. Then you flip the card over, noting whether you got it right or wrong.

With much practice, our brains can internalize certain algorithms. Each and every one of us does it all the time. Consider a task like placing brackets on teeth, something that thousands of orthodontists do every day at lightning speed. Bracket placement is highly complex—each type of tooth has its own optimal x,y location and angle at which to align the bracket, differing only by several half-millimeters and degrees. When starting off, student orthodontists are taught to look at a table specifying the measurements and to use gauges and protractors to place the brackets in the right spot. But you can't run a profitable practice that way. That's why, with much practice, many orthodontists eventually discard the gauges and do it by eyeballing. And they tend to get it very close, if not spot-on. Eventually, when they see an upper left central incisor, they can't help but think "4 millimeters from the incisal edge angled 12 degrees mesially" (if they're using a straight-wire prescription).

When an algorithm is internalized, it causes the brain to rewire itself. An algorithm for calculating large sums of numbers will develop local regions in the parietal cortex on the top of the head. Most algorithms require forming new associations between two concepts. Internalizing these new associations is neurologically similar to acquiring synesthesia.

To understand what goes on during brain rewiring, we need to look briefly at brain development in children. An infant's brain contains about eighty-six billion neurons. In contrast to what was previously believed, the brain has the capacity to generate new neurons throughout life. The hippocampus, the main control center for long-term memory, for example, continues to create new cells well into old age, and antidepressant SSRIs can facilitate the generation of new neurons in the hippocampus.

Although the brain *can* generate new neurons, the number of neurons stays relatively constant throughout life. However, the brain grows to become five times as big by the time of adulthood. How does the brain become that big, if the number of neurons stays constant? First of all, the brain generates additional synaptic connections between neurons. The brain rewires and fine-tunes its connections differently depending on the relative timing of sensory stimuli. Neurons that fire in synchrony form strong, stable connections, whereas neurons that fire out of synchrony will end up destabilizing and withdrawing connections.

McGill University psychologist Donald Olding Hebb first stated this in 1949 as a principle: cells that fire together wire together, whereas cells that fire out of sync lose their link. As an example of how neurons can come to fire and wire together, consider the well-known case of Pavlov's dogs. In 1901 the Russian psychologist Ivan Pavlov found that when a buzzer regularly sounded at the time at which he fed his dogs, the dogs would come to associate the sound with food and would begin to salivate prior to any appearance of food. This has come to be known as classical conditioning. What happens in this case is that neurons that process the sound from the buzzer and neurons that initiate the secretion of saliva become wired as a result of continually firing together.

Synchronicity here is key. When connected neurons fire simulta-neously, producing brain waves in the same frequency range, they are able to work as a unified whole for a period of time. This mechanism, which is also known as synchronous oscillation, has been proposed to account for basic aspects of perception such as sensory binding (that is, the process by which the information of color, movement, and form becomes integrated into unified perceptions, and into more gen-eral processes like consciousness and attention). But ongoing syn-chronicity is also important for maintaining wiring among neural networks. Once neurons "fire out of sync," the neural network is not maintained.

In a study published in the May 2014 issue of the journal *Science*, Edward Ruthazer from the Montreal Neurological Institute and Hos-pital and colleagues were able to confirm and expand on this theory by observing neurons in a microscope. They found that asynchronous firing not only caused neurons to lose their ability to make the recipi-ent cells fire, but also caused them to make up to 60 percent new branches in search of better-matched partners. So, at the time at which neurons lose preexisting connections, new brain connections are particularly likely to be formed.

There is a second factor that contributes to the growth of the brain that can be triggered even in late adulthood. Adult neurons have long fatty axons that are crucial to brain communication. My-elination of nerves involves adding of fat to the brain's axons, which increases brain size and makes neurons transmit information faster. The fat cells wrap around axons, which helps keep energy in the axon and makes the action potential (the firing of a neuron) go down the axon faster. But not all nerve cells are myelinated, and not all to the same degree. So, what determines the difference? One factor is a ba-sic energy or fuel molecule called adenosine that is released when a neuron fires. Adenosine binds to receptors on the myelin-making oligodendrocyte cells. This makes them start wrapping an axon in myelin.

An unmyelinated axon is akin to a July-4th sparkler: waves of electrical impulses travel down the axon at a rather dull rate. Myelination increases the speed and strength of the nerve impulses by forcing the electrical charge to jump across the myelin sheath to the next open spot on the axon, turning the electrical impulses into bottle rockets. Myelin thus allows us to send messages from one neuron to another neuron much more quickly than if it were not there. Normal nerve function is lost in demyelinating disorders, such as multiple sclerosis.

Myelin develops rapidly in children's brains and may underlie some of the fundamental changes children go through when they suddenly develop a new ability. But we continue to generate myelin after childhood, though not as quickly. When we practice associating one thing with another, like the idea of water with the Spanish word *aqua*, this generates myelin, making the neural connection faster and stronger. With enough of a certain type of practice, you trigger myelination and develop new skills.

Using diffusion tensor imaging (DTI), which measures myelin, or white matter connectivity, researchers have been able to show that professional piano players and jugglers have more myelinated axons in a certain area of the brain, and that the myelination correlates with the amount of time spent practicing. When practice changes the timing between neural circuits in various brain areas, this changes the way the circuits compute. While the brain keeps generating new neural connections all of your life, the greatest number of new neurons, new neural connections, and myelinations are made before adulthood.

During brain growth the brain doesn't just make new connections, it also gets rid of the connections it doesn't use. This process of trimming the neural connections is known as neuronal pruning, or just pruning. Pruning is thus a process that changes the neural structure by reducing the overall number of synapses. This results in more efficient synaptic configurations. Pruning is governed primarily by environmental factors, particularly learning.

The brain can also change its wiring in a different way. In the pruning process, neurons don't die off. They simply retract axons from synaptic connections that are not useful. But the brain can also rewire itself by killing off its neurons in a process that is called apoptosis, which is a form of programmed neuronal death different from the kind of killing of neurons that occurs in brain injuries. In apoptosis the neuron is killed and all connections associated with the neuron are also trimmed away.

During childhood and adolescence, initially imprecise, unused, or unnecessary connections between neurons are gradually pruned away, leaving connections that are stronger, more useful, and more specific. We can think of it as a sort of neuronal natural selection.

In some individuals, however, the pruning processes deviate from the norm. So, direct neural connections between regions of the brain that do not normally stay connected remain so. For example, connections may remain between the color and form areas, or between color and auditory brain regions. The former would lead to grapheme-to-color synesthesia and the latter to sound-to-color synesthesia.

Unconscious Shortcuts

With increased myelination, practiced associations eventually become automatic. This is what it is to internalize algorithms and synesthetic connections. Associating one thing with another is a shortcut that helps memory. In fact, the brain unconsciously makes use of algorithms and shortcuts whenever it can. Recall the case of the baseball outfielder from chapter 1. The outfielder's brain doesn't observe that the ball is traveling 100 miles per hour at a forty-five-degree angle and then tell the player's legs to run thirty feet to the right because that is where it has calculated the ball will land. Instead it uses a rule of thumb that adjusts his running speed so that the ball continually moves in a straight line in his field of vision, putting him exactly where he needs to be when the ball comes down. It's a shortcut.

The brain is wired for these shortcuts; they help the unconscious brain make fast calculations when needed. Another common shortcut is to rely on perceptual expectations. For example, the brain makes predictions about the whereabouts of moving objects to enable us to see them. While the brain can unconsciously calculate the speed and position of very fast objects, it cannot do this in real time. When human eyes see an object like a speeding car, it takes one tenth of a second for the brain to process that information. So, information that the brain receives from the eyes is already outdated by the time it reaches the visual cortex. To compensate for this, the brain makes estimations about where the object is going to be in the near future. It thus tracks moving objects as being farther along in their trajectory than what a person actually sees with their eyes. In this way, we can bypass the delay in processing time.

Predicting ahead for just long enough to track the environment correctly is a special case of a general shortcut that drives visual perception. The imprint the environment makes on the retina in the eye is basically two-dimensional and nondynamic. To get from a two-dimensional and nondynamic retinal imprint to a three-dimensional dynamic scene, we must rely on expectation. What the brain expects us to see influences our perception of what we actually do see. Of course, when the world is not as expected, we don't see the world as it is. So, we are susceptible to visual illusions. A good example of this phenomenon is the Müller-Lyer illusion.

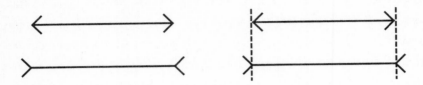

The Müller-Lyer Illusion. Even when you learn that the line segments on the left have the same length, they continue to appear as though they are different lengths.

The direction of the arrowheads at the ends of the lines affects one's perceptual experience: a line appears shorter when the arrowheads are turned inward, but longer when they are turned outward. The illusion persists even when we come to believe that the lines have the same length. We only see the lines as having the same length when we add vertical lines that allow us to compare them.

There are several possible explanations for why the Müller-Lyer illusion works. The most popular explanation is based on depth perception. Depth perception involves generating an internal three-dimensional model of the environment. Part of the mechanism that produces the three-dimensional model adjusts for the sameness in size of objects located at different distances from us. This is also known as size constancy. This mechanism ensures that objects are not perceived as shrinking when we move away from them. As a result of this process, the line with the outward hashes appears as if it were actually closer than the other line, and thus the line seems longer.

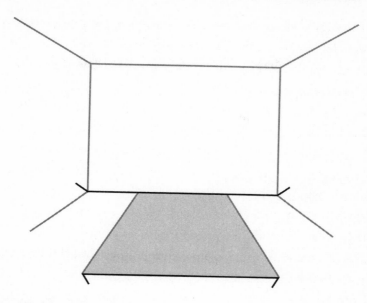

The Müller-Lyer Illusion. Illustration showing how outside corners generate the appearance of an object being farther away from us, whereas inside corners make the object appear closer to us.

So why do we need shortcuts? One reason is speed in reasoning, calculation, and decision-making. Another is that unconscious short-cuts can minimize errors that our conscious mind makes. To see how this works, complete the following five mental exercises.

Exercise One: Count the number of *f*'s in the following sentence:

Finished files are the result of years of scientific study combined with the experienced number of years.

How many *f*'s were there? Six. If you are like most people, you counted three. This is because we tend to ignore the high-frequency word *of*.

Exercise Two: A bat and ball cost a dollar and ten cents. The bat costs a dollar more than the ball. How much does the ball cost?

The vast majority of people quickly and confidently respond with "ten cents." But this is incorrect. The correct answer is one dollar and five cents for the bat and five cents for the ball. If the ball had cost ten cents and the bat costs a dollar, then the bat would have cost only ninety cents more than the ball.

Exercise Three: If you pass the second person in a race what place will you be coming in?

Most people reply "first." But if you pass the second person in a race, you still need to pass the first to come in first. So, the correct answer is second.

Exercise Four: If a plane crashed on the border of England and Scotland, where would they bury the survivors?

People tend to respond with "their country of origin." The true answer is that you don't bury survivors. That would count as murder.

Exercise Five: Based on a conversation with your new acquaintance Christina you discover that she has a son. You subsequently find out that she has two children. What is the probability that they are both boys?

Assuming that any birth is equally likely to be a boy or girl, most people answer 1/2.

In fact, the correct answer is 1/3. The important thing to note is that the problem is not asking for the probability of an individual child being a boy. It is asking for the probability that *both* children are boys—that the first child is a boy *and* the second child is a boy. In a family with two children there are four possible combinations of boys and girls:

First child	Second child
Boy	Boy
Boy	Girl
Girl	Boy
Girl	Girl

In this particular case we know that the fourth combination (Girl, Girl) is impossible because at least one child is a boy. The remaining combinations are equally likely and so each has a probability of 1/3. Only one of these leads to both children being boys. And so the probability that both of Christina's children are boys is 1/3, while the probability that one is a boy and one is a girl is 2/3.

Solving word and number problems is a task for our conscious mind, which makes mistakes when it uses shortcuts, whereas planning for complicated physical actions like coordinating the movement necessary to catch a ball or pick up a cup is a task for our unconscious mind, which also uses shortcuts yet is not so error-prone. If you unconsciously estimate that you are reaching to and grasping a mug full of coffee and it's actually empty, the mug may almost fly out of your hand, because your brain calculated the effort needed on the basis of an incorrect expectation of weight. But this is a mistake in your observation about your environment and not in your brain's internalized calculation.

Conscious thought has its benefits. It makes us access information and weigh options rationally and act according to a carefully executed plan. But that comes at a price. Conscious processes and reasoning in the realm of calculation are riskier and more frequently mistaken than unconscious brain processes. That is not to say that implicit biases and cognitive shortcuts (heuristics) cannot produce equal or greater errors. But in exercises of pure counting and calculating, the conscious brain tends to make more mistakes than the unconscious brain. To illustrate, consider the Ebbinghaus illusion (also known as Titchener circles), which consists of a change in the perceived size of a circle surrounded by circles with a larger or smaller area:

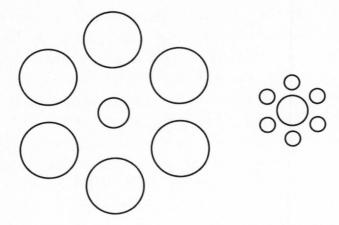

The Ebbinghaus Illusion. Studies have shown that this illusion leads to a misperception of the size of the central circle, but only marginally (if at all) affects grasping behavior directed at the central circle.

The illusion is an excellent demonstration of the fact that the conscious perception of an object depends to some extent on what surrounds it. Psychologists Dominic W. Massaro and Norman H. Anderson have proposed that the Ebbinghaus illusion occurs owing to a cognitive mechanism of size contrast, which alters the consciously perceived size of an object, adjusting its relative smallness or largeness relative to the objects that surround it. One reason for this effect is that larger objects tend to pop out and grab our attention just like a differently colored object placed among a crowd of same-colored objects would do, thus placing the smaller objects in the background, which makes them look even smaller than they are. This is likely a convenient adaptative mechanism when we want to consciously recognize an object and consciously separate it from its surroundings.

Although we cannot normally help but consciously see the inner circle as differently sized in the left and right figures, it turns out that our unconscious visual system usually does not make this mistake. In studies where participants were asked to reach to and grasp the circle in the middle, their hand aperture was exactly the same in the two cases. So, our conscious vision for our perceptual recognition system and our reasoning system are more susceptible to error than our fast, automatically calculating unconscious brain.

But if conscious perception and reasoning introduce errors, why then do we need consciousness in the first place? Well, it could be that we don't really *need* it but that consciousness is something that has evolved as a side effect of something else we needed for adaptive purposes. Our conscious, slower perceptual and cognitive systems are associated with long-term planning of action. The fast unconscious systems are always "on the go." They don't typically store information and they are not useful for planning for the future. Planning the future is an important part of survival and flourishing. So, our slower conscious perceptual and cognitive systems seem to serve an important role in our lives, despite their susceptibility to error.

If our brains can learn algorithms and come to resemble the brains of autistic savants, why can't we all become, say, bracket placement experts or professional outfielders overnight? Because our brains haven't internalized the algorithm required to complete the task. Training for ten thousand hours is one way to achieve that result. But by purposely internalizing algorithms, we can speed up the process. We just have to supplement the brain's normal ways of functioning so it "makes room" for the extra algorithms, changing its neural structure in the process. How do we do that? We've already shown that training isn't the only way to gain extraordinary cognitive talent, but that doesn't mean that wannabe savants should throw in the towel. Training in conjunction with the right algorithms can sometimes unlock the brain's dormant abilities. This is the case for many people participating in memory sports.

CHAPTER 4

What's Your Number?

*Memory Sports and Reciting Pi to the
60,000th Decimal Point*

As a baby, Kim Peek, who was the real-life inspiration for Barry Levinson's 1988 movie *Rain Man*, was diagnosed with mental retardation, and physicians told his parents that he would never be able to read or talk. They recommended sending the little boy to a mental institution and getting on with their lives. But Kim's parents chose to raise him at home. They quickly realized that the boy with the oversized head had a remarkable brain. Due to his parents' efforts, Kim had the opportunity to develop his amazing talents. A large head does not equal intelligence or better brain processing. But it does provide more storage space for someone who is able to process the content of ten thousand books, which was the number Peek had read by the time of his death in 2009.

Kim Peek's special abilities started early, around the age of a year and a half. He could read both pages of an open book at once, one page with one eye and the other with the other eye. He apparently had highly developed language areas in both hemispheres. He used this style of reading for his entire life. His reading comprehension was impressive, and he could retain 98 percent of the information he read. Since he spent most of his days in the library with his dad, he quickly made it through thousands of books, encyclopedias, and maps. He could read a thick book in an hour and remember just about anything in it. Because he could quickly absorb loads of information and recall it when necessary, his condition made him a living encyclopedia and a walking GPS. He could provide driving directions between almost any two cities in the world. He could also do calendar calculations ("which day of the week was June 15, 1632?") and remember old baseball scores and a vast number of musical, historical, and political facts. His memory abilities were astounding. Kim Peek was a true super-savant.

Most people with savant syndrome suffer from autism. However, in spite of lending inspiration to a movie character with the condition, Kim Peek was not himself afflicted with autistic spectrum disorder. The main cause of his remarkable abilities seems to have been the lack of connections between his brain's two hemispheres. An MRI scan revealed an absence of the corpus callosum, the anterior commissure, and the hippocampal commissure, the three main parts of the neurological system that transfers information between the right and left hemispheres of the brain.

Kim Peek's condition, which is known as agenesis of the corpus callosum, is among the most common human brain malformations, occurring in at least one in four thousand individuals. Normally, a rudimentary form of the corpus callosum appears during the twelfth week of pregnancy, and continues to develop during pregnancy and early childhood. In fact, this important nerve bundle is still maturing until early adulthood. There are numerous factors that can cause an absence or a malformation of the structure, including genetics, infections, and intoxication during pregnancy or early childhood.

Individuals without an intact corpus callosum sometimes experience linguistic and social impairments. The main linguistic and social problems stem from difficulties understanding nonliteral language, including idioms, proverbs, irony, sarcasm, subtle jokes, and conversational implications. Most adult English speakers will understand that "all good things come to he who waits" is advocating patience and is not to be understood literally. But for a person without a fully developed connection between the two hemispheres, the proverb may stand out as a plainly false statement saying that if only you wait, you will get everything that is good. The corpus callosum apparently is required to be able to override the literal meaning of language and rely on context for interpretation.

The lack of a connection between the two hemispheres also seems to promote a greater focus on, and interest in, details and patterns and a need for rigorous routines and repetition. New research shows that the lack of a corpus callosum, while inhibiting communication between brain hemispheres, may actually lead to increased brain connectivity *within* each of the two hemispheres. This increased local connectivity may explain both the tendency to be obsessed with details and the increased intellectual abilities of people like Kim Peek. The fact that the lack of the major connections between the two hemispheres gives rise to autism-like symptoms may also explain why people like Kim Peek often are thought to be autistic.

Split Brains

In some sense Kim Peek was a natural born split-brain patient. The neurological bridges between his brain hemispheres simply never developed. But in the late 1960s, split-brain surgery was introduced as treatment for patients with intractable epileptic seizures. The surgical procedure involves severing the corpus callosum, the main fibrous bond between the brain's left and right hemispheres. This helps with the seizures in most cases. But after a split-brain surgery the two hemispheres do not exchange information nearly as efficiently as before, which sometimes gives rise to changes in behavior and agency.

Psychologists Michael Gazzaniga and Roger W. Sperry, the first to study split brains in humans, found that several patients who had undergone a complete removal of the corpus callosum suffered from split-brain syndrome. In these patients, the impulsive right hemisphere, which controls the left side of the body, acts independently of the logical and rational left hemisphere, which controls the right side of the body. This can give rise to a kind of split personality, in which the left hemisphere gives orders that reflect the person's rational goal, whereas the right hemisphere issues conflicting demands that reveal hidden preferences. One of Gazzaniga and Sperry's patients pulled down his pants with the left hand and back up with the right in a continuing struggle. On a different occasion, this same patient's left hand made an attempt to strike his unsuspecting wife as the right hand grabbed the villainous limb to stop it.

Another of their patients, Paul, had a fully functional language center in each hemisphere. While the hemispheres control the opposite sides of the body, information from each eye normally enters the same hemisphere in which it's located and gets transferred to the other hemisphere through the corpus callosum. But this transfer was significantly inhibited in Paul's brain, making it possible for the researchers to question his left and right hemispheres independently. They started by blindfolding Paul's left eye while having him read text with the right eye. In one case, they asked the right side what their patient wanted to be when he grew up. With his left hand, he wrote, "an automobile racer." When they reversed the experiment, posing the same question to the left brain, however, he responded "a draftsman."

Kim Peek is similar to Paul in this respect. There is no doubt that he must have had a fully developed language center in both hemispheres. Language is normally processed in areas of the temporal lobe on the left side of the head. When you read with your left eye, the information first ends up in the right hemisphere. This is because the right side of the brain receives the information from the left half of the visual field. For example, when you look at an open book, the left page is in the left half of your visual field, while the right page is in the right half of your visual field. So, your visual experience of the left page is in the right side of your brain, while your visual experience of the right page is in the left side of your brain. Since the language center is in the left side, the information from your left eye must be transferred back to the left hemisphere, via the corpus callosum, to be processed by the language center. This long transfer from one side of the brain to the other is usually a disadvantage. But Kim Peek didn't have a corpus callosum or a hippocampal commissure. So, his brain would have had to develop the abilities to process language in both hemispheres. This, of course, gave him a major advantage in terms of speed-reading and information retention.

You might think the same would apply to other hemisphere-specific abilities, such as visual imagery and math, which are primarily left-hemisphere based. However, Kim Peek was unable to "reason his way through" mathematical problems. Despite his brilliant mind, his IQ was 87, significantly below normal. It was also difficult for him to follow instructions that required any sort of metaphorical interpretation. One time, Kim's father asked him to "lower his voice" at a restaurant, only to observe Kim sliding down into his chair, lowering his voice box in the process.

But there are several respects in which Kim Peek was not like Gazzaniga and Sperry's split-brain patients. He did not exhibit any symptoms of truly split personality or conflicting desires deriving from separate hemispheres. While surgical severing of the brain hemispheres normally results in a disconnection syndrome, individuals born without a corpus callosum and patients who undergo the surgery as young children do not suffer these consequences, suggesting that the brain of a fetus or small child is plastic enough to develop alternative neural pathways for this kind of information transfer.

In Kim Peek's case, information that didn't need to travel simply stayed put in its respective hemisphere. But in some situations, he was able to combine information from both hemispheres to yield a whole. He could give a full account of any book he read, as opposed to giving two separate accounts of all the left-hand pages and all the right-hand pages. So how did Peek avoid this split in information integration even though he lacked the three main connections between the hemispheres?

A likely answer is that information was able to travel through subcortical connections in the brain's ancient regions. While the brain transfers most information between hemispheres via the corpus callosum, we know that certain kinds of information, such as physical pain, are transferred through subcortical areas. Normally, it is a relatively small amount of information that is transferred that way. But Kim Peek may have developed additional subcortical connections for information transfer. That is, his brain may have relied on older pathways more deeply embedded in the brain for getting information from one hemisphere to the other. This is exactly what happens in nonhuman animals that don't have a corpus callosum, such as kangaroos, wallabies, and possums.

Acquiring Kim Peek's Memory Abilities

Can those of us without split brains acquire amazing memory abilities like Kim Peek's? It seems that we can. Mark Aarøe Nissen is an average twenty-six-year-old student at Aarhus University, Denmark, except for his extraordinary memory abilities. The math major has competed in memory sports for several years. He can recite pi to more than twenty thousand decimal points, recall thousands of names, faces, and historical dates, and keep track of four decks of cards while figuring out the best bidding strategy at the casino.

For those of us who can barely remember the items on a grocery list, these extraordinary memory skills might make him seem like he was born with savant abilities. But Mark laughs as we make this suggestion. "That's nonsense," he insists. "It's not something that just comes to me. I train to accomplish what I do."

Mark's training consists in part in developing mnemonic devices. Mnemonic devices are strategies, such as acronyms, that can assist memory. You are probably familiar with Roy G. Biv, an acronym for the sequence of hues commonly described as making up a rainbow: Red, Orange, Yellow, Green, Blue, Indigo, and Violet. The acronym can help you recall the number and order of colors. Mnemonic devices rely on our short-term memory to execute each step in the process. Short-term memory is also known as working memory. It holds information you are actively using at the moment. For example, if you need to address an envelope, you would need to hold the address in your working memory long enough for you to write it on the front of the envelope.

Short-term memory is limited in its capacity. In his seminal 1956 paper, Princeton cognitive psychologist George A. Miller argued that memory is limited to nine items. That's hardly more than the digits in an American license plate number, and it fades quickly. As you're desperately trying to hold on to the number of that hit-and-run driver, working memory is waiting on something more pressing to replace it with.

So what benefits does working memory offer? It's fast. Its speed is what makes it particularly useful for manipulating information we're conscious of, like reasoning and calculation. In fact, the speed of your particular working memory can sometimes mean the difference between a high-functioning brain and a learning disability.

When Mark competes in memory sports he uses a headset that blocks out sounds. Some use special glasses to block out peripheral visual input but Mark doesn't find that necessary. He naturally goes into his own world of emotions, shapes, and landscapes. Mark has once attempted to beat the current European record in reciting pi to the greatest decimal point. The record is currently held by Daniel Tammet, the author of the bestseller *Born on a Blue Day: Inside the Mind of an Extraordinary Autistic Savant*. Tammet won the European championship with 22,514 decimal points, on International Pi Day (March 14) in 2004.

Regrettably, Mark didn't succeed. After reaching the 17,108th decimal point, he wrote 48 where he should have entered 84. He immediately realized that he had made a mistake and stopped the process. We asked Mark why he made the mistake. "Concentration," he replied. "After hours and hours of writing down digits of pi, your brain just stops focusing. It's a bit like hitting the wall in a marathon." The infamous wall occurs around the twenty-mile point as a physiological response to carbohydrate depletion. Most new marathon runners have heard of it but don't expect it: "The wall. Yeah—well—if I see one I will break through it or climb over it," they say. Then at about twenty miles they start to realize what it really means to hit the wall. The legs start tingling until they become almost numb. Then the pain sets in. Red hot lead inside the legs. For Mark it wasn't quite that bad. But it only takes a fraction of a second to lose concentration and screw up.

Both Mark and Tammet have some way to go before they reach the superhuman abilities of Chao Lu, who holds the Guinness world record in reciting pi dating back to 2005. Lu recalled 67,890 digits in 24 hours and 4 minutes, ending with an error at the 67,891st digit: a 5 where there should have been a 0. Most of us have experienced fatigue after a few hours of standardized testing. It's a bit difficult to contemplate the mental and physical fatigue Lu must have gone through during this daylong recall exercise.

Like Mark, most of the people in memory sports use only mnemonic techniques and their seemingly neurotypical brains to achieve their goals. The mnemonics they use are specifically designed to make use of cross-sensory processing that is similar to that seen in synesthesia.

Some mnemonics used to aid memory rely heavily on narratives. Our brains love stories, so much so that it's much easier to retrieve the extensive details of an event that occurred many years ago than it is to remember the sixteen digits of a credit card number you look at every day.

Our brains also love emotional intensity, which helps to explain why some aspects of early childhood survive childhood amnesia. Setting in around age seven, childhood amnesia involves the sudden deletion of previous memories. It was hitherto thought that this amnesia occurred because the brains of young children couldn't form lasting memories of specific events. But in the 1980s, Patricia Bauer—a professor of psychology at Emory University—and colleagues started testing the memories of children as young as nine months old. They discovered that children have very solid long-lasting memories of specific events. But as the children grew older, it seemed that they somehow lost these memories at some point. Bauer and her team recorded children at the age of three talking to a parent about previous events, such as going to Six Flags or on a vacation to the US Virgin Islands. At the age of seven, these children could still recall more than 60 percent of the recorded events, but children who were just a year older remembered only about 40 percent. Age seven seems to mark the onset of childhood amnesia. The reason childhood amnesia sets in around that time has to do with pruning, the main purpose of which is to get rid of unused or ineffective brain connections. Many of our early memories are pruned away, but as pruning is radically reduced in adulthood we have better memories of specific events as adults. Some childhood memories survive this pruning, and they tend to be very emotional or strongly connected to a story with a very intense plot.

Knowing what the brain is fond of, we can use it our advantage. Rather than trying to remember numbers as numbers or cards as cards, memory champions generate meaningful associations between the things they want to remember, like numbers, and things they already remember, like objects in their homes, people they know, and emotionally intense events from the past.

Mark's strategy in memorizing long strings of numbers is to associate each set of four digits with a particular object that plays a role in a narrative. So recalling a sixteen-digit number is as simple as remembering four particular objects in the correct order. These associations, which begin as constructions, soon become so ingrained that they end up being completely automatic, thus mirroring the sort of process we see in synesthetes. By relying on these mnemonics, people who train for memory competitions thus acquire a form of synesthesia that can help them store and retrieve large sums of information. Virtually anyone can become a memory champion. It just takes practice.

When Mark first began practicing to remember pi, he broke the number up into a series of two-digit numbers and converted them into shapes. He used shapes of objects from the real world: humans, animals, and fruits. "The number thirteen used to be the shape of an apple and thirty-two used to be fire," Mark explains. Once he had memorized the shapes, he planted them in imaginary but familiar landscapes:

his childhood home, the university, his hometown, that vacation spot on the south coast of France. Today Mark remembers exactly ten thousand shapes, each representing a four-digit number. When he needs to recall a number he walks through the landscape and sees the shape, and the number pops into his mind. He thus uses these shapes to quickly memorize new strings of digits. Finding his way through the landscape is simple, he says. "It's just like following a route you know, for instance, the route from your home to your workplace."

To illustrate the method, imagine you need to memorize a series of digits: 3142591118. We might divide it up as follows: 314-25-911-18. The first three digits are the first digits of the number pi, and the next two digits are the date in December on which most people celebrate Christmas. The date of the most talked about terrorist event in the United States is 9/11, and 18 is the legal age of majority in the United States.

You now need to arrange these associations in a particular order, so you don't end up recalling the number as, say, 2518314911. Here is where narrative enters the picture. For example, you might imagine sitting in a math class in college learning about pi (314) just before Christmas (25) when you hear about a new terrorist attack (911), this time committed by a minor, who is under 18.

By practicing the story in your head while trying to visualize the stages of the story, the associations and their particular order end up becoming relatively automatic, which means that you don't have to go over the story in detail to retrieve the digits. Emotionally charged memories can assist in automatizing the connections. For example, the teenage terrorist in charge might be your son or an ex who broke up with you in the meanest of ways.

When testing the speed of this method I found that I could acquire around two thousand digits of the number pi during a flight from Edinburgh to New York. In my own experience, it is better to use multiple two-hundred-digit narratives rather than one long narrative in which you are more likely to get lost. In my own case, my first narrative depicts the events of one of the first days during my postdoc in Australia. I started with the five digits that virtually everyone knows, namely 31415. As I entered the Coombs Building, I met a guard who looked like he was 92. I then went to my office and continued working on an article for a Festschrift, a tribute in writing to an academic colleague. People tend to be 65 when they receive Festschrifts in their honor. I then talked to a student, who was around 35 at the time. Then I went to the secretary Di's office. Di's mother was there. I formed the number associations Di's mother-89 and Di-79. Then I went to afternoon tea, which rhymes with 323. As I returned to my office, I finished reading George Orwell's novel *1984*, shortening it to '84 in my mind. I then called home. 6264 reminds me of my parents' phone number. This marks about one half of my first narrative and the first twenty-four digits of pi: 3.141592653589793238846264.

By using this method but substituting your own number associations and narratives, you can learn to remember between one hundred and three hundred digits in about one hour. To beat the European record you would need to spend no more than 240 hours practicing: much, much less than the ten thousand hours that is alleged to be the golden number for acquiring an expert skill.

Acquiring synesthetic number-color associations before attempting to beat the European record in reciting pi is likely to help you progress. For example, if 9 is red, 2 is yellow, and 6 is green, you can retrieve the digits 926 when seeing the red, yellow, and green as you stop at a traffic light. We turn to how to learn to be a synesthete next.

How to Become a Synesthete

Synesthesia can aid memory. But can you really learn to become a synesthete? Neuroscientists Nicolas Rothen and Beat Meier have argued that it's not synesthesia per se that enhances memory but rather the fact that synesthesia can be used as a mnemonic strategy consciously or unconsciously. It's a mental aid in learning and memorizing. Although most cases of synesthesia are present either at birth or in early childhood, or acquired in strange ways, synesthesia is not reserved for the lucky few. Because synesthesia is an extreme variant of multisensory processing, it is possible for all of us to obtain the intellectual and creative advantages of the condition by reinforcing naturally occurring multisensory processes and by following the lead of synesthetes in our pursuit of originality, innovation, and human understanding.

Researchers at the University of Amsterdam have looked into the question of whether synesthesia can be learned. Their findings, which are published in the June 2012 issue of the journal *PLOS ONE*, show that on the basis of a simple two-week training, what they call quasi-synesthesia can be learned.

The researchers focused on grapheme-to-color synesthesia. Seventeen nonsynesthetes were asked which colors they preferred for the letters *a*, *e*, *s*, and *t*. These preferences, of course, were more or less arbitrary. They were then given specially printed books, with *a*, *e*, *s*, and *t* printed in colors, according to their preference. Each participant read, on average, 100,000 words of customized, colored text. To their surprise, the researchers found that it seemed to take only around 49,000 words to create a Stroop effect in participants.

You might have encountered the Stroop effect before. It often happens when you see the name of a color, such as *red*, printed in ink of some other color, such as green. If you're asked to name the color of the *ink*, it takes significantly longer for you to do it if the ink color doesn't match the word.

RED RED

The word *red* is here displayed in the color black (left) and the color gray (right: representing green). It takes longer for subjects to read the word *red* when it is printed in green than when it is printed in black or red.

The researchers' findings suggest that colored-letter exposure can strengthen color-letter associations. This, of course, is not at all surprising given that colors are very memorable. As we will see in chapter 5, color-number associations also undergird the Universal Concept of Mental Arithmetic system that teaches children to perform highly complex calculations mentally.

None of the research participants experienced a "pop-out" effect. This effect is typically associated with a form of synesthesia known as projector synesthesia, a condition where the synesthetic experience seems to be projected out into the external world. The first "pop-out" study was conducted by neuroscientists Vilayanur S. Ramachandran and Edward Hubbard. The study tested two synesthetes as well as neurotypical controls with an array of synesthetic color-inducing graphemes. Within each array, target graphemes were arranged so that they could be grouped together into simple shapes.

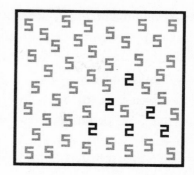

When normal subjects are presented with the figure on the left, it takes them several seconds to identify the hidden shape. Some grapheme-to-color synesthetes purportedly can quickly recognize the triangular shape because they experience the numerals 2 and 5 as having different colors.

Participants were presented with each array for the duration of one second, and then asked to name the correct shape from a group of four alternatives. Their findings showed that some synesthetes have higher accuracy or quicker reaction times in determining the hidden shapes compared to the ordinary population. The fact that none of the volunteers in the learning study experienced a pop-out effect indicates that they likely became associator synesthetes—the more common form of naturally occurring synesthesia. Associators see synesthetic experiences only in their mind's eye, or simply report knowing that one thing is associated with another, unlike projectors, who see them as projected out into the external world.

In our lab we wanted to go one step further. Individuals who become extremely talented in what are called "memory sports" use emotion as part of their system for remembering random sequences of numbers or long lists of names. The reason these champions typically use emotions to generate stable memories is that memories attached to emotional content are more vivid and long-lasting than memories that are not attached to the brain's emotional system. It is widely accepted that the most vivid autobiographical memories tend to be of emotional events. That is, we remember emotionally charged events better than boring ones.

Neuroscientifically, it is a little more complicated than that. Emotions are characterized on a spectrum of arousal—whether they are calming and soothing or exciting and agitating—and valence—whether they are positive or negative. The arousal dimension of emotion as well as extremeness of valence are critical factors contributing to the enhancement of memory.

Making use of the knowledge we acquired from memory sport champions, we thus decided to associate letters and numbers not just with colors but also with emotions. To generate an emotional memory associated with a color and a letter or number we used emotionally disturbing or interesting quotations as the colored reading material for the participants.

We began with one letter at a time, and added a new letter each day. Below, we will illustrate the program using the first four letters of the alphabet.

In this example, the learner has a preference of red for *a*, white-on-gray for *b*, blue for *c*, and white-on-black for *d*. These colors, along with the others in the system, are chosen because lots of people happen to share these particular color preferences. It is easier to strengthen preexisting symbolic associations between colors and emotions than to generate new ones. This is also called color symbolism: the conscious or unconscious associations that we are conditioned to make by our environment.

Colors are already laden with symbolism in our minds. Red is the color of blood and has associations with war and aggression but also love. When someone is angry they "see red." Red also signals "stop" as a result of traffic signs. In the Western world yellow means caution as a result of street warnings of upcoming curves, pedestrian crossings, and animal crossings. When someone is seen to be afraid they are sometimes called yellow. When someone is feeling down or depressed, they are said to be feeling blue. Environmentally conscious companies are said to be "going green." When something is considered borderline, it is a "gray area." White represents virginity, purity, and the body of Christ. Black represents death or sin in Catholic liturgy. Red, white, and blue symbolize patriotism in America. Red and green symbolize Christmastime. Black and orange are associated with Halloween.

There are also significant cultural variations in color symbolism. In Islam green is the sacred color. In Ireland green is considered lucky. In Maharashtra, India, it represents life and happiness. In Europe purple is associated with royalty because until relatively recently it was an extremely expensive dye and only royalty could afford it. In Catholicism it is the liturgical color for the seasons of Advent and Lent. In Hinduism, saffron is the most sacred color, representing fire that burns impurities. Buddhist monks wear saffron robes because it used to be the least expensive color dye, symbolizing simplicity and detachment from materialism. In the Native American Navajo Nation turquoise, white, yellow, and black represent four sacred mountains. In the Apache Nation green, white, yellow, and black are sacred colors, and in the Iowa Nation the sacred colors are black, yellow, red, and white. There is significant individual variation as well. You might have a special preference for blue walls, because your childhood bedroom had blue walls.

General instructions: The sentences are available at our lab page at brogaardlab.com/superhuman. Read the sentences for each day slowly one at a time every hour. Think about what the quotations mean. Then reread all the sentences from the previous days. This should only take about six minutes on average every hour, less at first and more later. The total amount of work adds up to an average of 96 minutes a day. To complete this in two weeks, change to half-hour intervals and do it twice a day with eight waking hours in between. At the end of the two weeks, read over all the sentences a few times a week to retain the synesthetic connections and foster stronger neural connections.

Not all of the volunteers who go through the training acquire synesthetic abilities. However, when we question people afterward, a consistent trend is that those who fail haven't put in the hours. Acquiring synesthetic abilities doesn't require Gladwell's ten thousand hours, but without consistent training every day the skills are going to be difficult to acquire. People who follow the routine acquire the associations to varying degrees. The vast majority pass the Stroop interference test described earlier in this chapter. Some begin to see letters printed in black ink as having colors, the way synesthetes do, although for most the grapheme-color associations are not very vivid visually.

In our latest testing round, nine out of the ten compliant volunteers passed the Stroop test. This result is similar to what we saw in the Amsterdam study cited above.

What is far more interesting is that six out of the ten compliant volunteers scored as synesthetes on the Eagleman Synesthesia Battery, which requires subjects to precisely and consistently identify the shades they associate with colors and numbers—this is a feat that nonsynesthetes cannot accomplish. Three additional volunteers scored as quasi-synesthetes (they came close to picking the same colors each time, but showed slight variations).

None of the ten compliant volunteers experienced pop-out effects in visual search. This suggests they likely became associator synesthetes.

Card Counting

Though our memory sportsman Mark can recite pi to the 20,000th decimal point, his favorite memory discipline is not reciting pi but counting cards. "I can easily beat blackjack," he says. "It mainly requires counting cards and then bidding in the right way. It's not that hard." When Mark told us this, we both immediately tried to convince him to join us at some casinos to make some quick cash. He didn't go for it. "For me, this is a sport, not a way to make money," he replied sternly. It seems that this memory superhero will only use his powers for good.

Realizing we weren't going to get Mark to join us on a *tour de casinos*, we wanted to know whether we could pick up these skills ourselves. Despite how difficult *Bringing Down the House* might make it seem, it turns out that card counting requires more social prowess than brute mathematical ability. In fact, you can learn to count cards with a few hours of practice, although you might need to practice a couple hundred more to make sure you don't catch the eyes of the pit boss. Card counting capitalizes on our natural proclivity to make synesthetic-like connections between pictures and numbers. Only rather than helping us keep a string of digits in memory, these connections help us remember all of the cards the dealer rapidly drops in front of our tablemates.

A typical casino game of blackjack utilizes between one and eight standard decks of fifty-two playing cards. The player's goal is to get the sum of the face values of all of her cards closer to twenty-one than the dealer, without going over. If the player fails to get closer to twenty-one than the house or if she goes over twenty-one, she loses all the money she bet for that round. So the basic strategy is to take enough cards to be somewhat close to twenty-one without taking so many you're forced over. But since there are a finite number of cards in each deck, there are only so many ways a deck can be ordered. As a simple example, if six players at a table using three decks of cards each have two kings in their hands, there's no possibility that the next card being dealt is a king. Likewise, if no kings have been dealt, then you're much more likely to be dealt a king than in the former scenario.

Card counting uses statistical probabilities to make more informed choices when placing a bet. Thus you can learn to predict what sorts of cards you're likely to get using very simple math.

The most proven strategy for counting cards is called the Hi-Lo method. High cards (aces, 10s, and face cards) are given the value of –1 and low cards (2–6) are given the value of +1. The remaining cards (7–9) are given a value of 0. All you have to do is add the cards up as you see them, starting at zero. For example, if the first twelve cards to be dealt are 3, A, K, 7, J, 2, 4, Q, 2, J, 8, 8, you would count them as +1, –1, –1, 0, –1, +1, +1, –1, +1, –1, 0, and 0. So the deck's sum works out to be –1. That number indicates that the house has the upper hand. Higher numbers are better than lower ones. In other words, you might want to hold off on upping your bet.

It's no secret that counting cards can get tedious. You'll have to sit down for a while if card counting is to work in your favor because keeping a tally of the deck's total value requires that you keep track of the cards from the beginning of each game. And since multiple decks are often used, you'll have to account for that number by dividing the running count by the number of decks in play.

Real skill is needed when you count cards in a fast-paced environment like a casino. Like any other billion-dollar industry, casino operators invest millions of dollars every year to make sure they maintain the advantage. So while counting cards may not be illegal, it definitely is frowned upon. Casino dealers, the floor staff, and the eyes in the sky are all trained to pick up the most obvious sign of card counting: someone concentrating on the cards like a zombie. To beat the house, you really have to get good enough that you can add new cards to your tally all while carrying on a casual conversation, sipping a cocktail, and cheering as everyone's favorite sports team makes a goal. Seem easy? With a little practice and your superhuman mind, it can be. We say "can" because counting cards doesn't supplant general blackjack strategy. It only lets you know when you have the edge. If you want to make a fortune, you still need to get good at playing the game. But the fact is you can make it to MIT level. After all, those kids never did as well as the movie made it seem.

The Name Game

We can all relate to that embarrassing moment at a party when you suddenly realize you don't remember the name of the person you've been talking to for ten minutes. It's the basis for one of the most memorable scenes in *The Devil Wears Prada*. Protagonist Andrea (played by Anne Hathaway) must memorize two books' worth of faces and names so that she can feed them to her impossible boss Miranda (played by Meryl Streep) at a Parisian gala. Is Miranda so pretentious that she cannot even be bothered to remember them herself? Or is there more going on? In our opinion, a little of both.

It turns out that of all types of words, names are the most difficult to get right. We often first learn someone's name when there are too many other distractions—the way they look, the topic of conversation, and how we look to them. It's not surprising that we might ask their name, yet not be able to recall it when put on the spot. This is called the tip-of-the-tongue phenomenon (TOT). Ever been in the middle of a heated discussion and not able to recall a word you used a hundred times the day before? TOT occurs not only with names, but in many other linguistic situations such as these. Research suggests that TOT occurs most frequently with names. In one study, 130 adults were asked to record every time they experienced the phenomenon over four weeks. The great majority of TOT cases turned out to occur for names of people the participants were well acquainted with, but hadn't seen in a while. So TOT is at least partially attributable to the regularity with which the name is used.

But the reason proper names are so hard to remember is that, unlike other words, names themselves have little to do with the individuals they are naming. Words like *river* or *snake* have concrete meanings. But for the average Joe, names like *Sarah* or *Justin* aren't connected with anything other than the people who happened to be given those names. It doesn't matter how unique the names are. In fact, it's just the opposite—you're more likely to remember a name if it's *not* so unique. This is because the more unique a name is, the more work your brain has to do to remember who it belongs to. For example, if you meet someone named Kahler, not only do you have to remember this name you've never heard before, you also have to remember that it belongs to the person standing in front of you. This explains why you might have trouble remembering Kahler's name, even though you might remember she's a law student from San Antonio. The only thing just as hard as remembering her name is remembering her phone number and address.

Names are best remembered when they're meaningful, but we already learned they don't have meaning. That's okay, you can assign it. Research shows that the mere act of making some connection between a newly learned name and some detail about the person or even someone else with the same name helps you remember the name. The first step is to remember to care. Seriously. Sometimes you can get so caught up with meeting someone that you forget to listen to them tell you their name. But once you master the art of listening, there are a few more tactics you can use. Benjamin Levy, author of *Remember Every Name Every Time*, developed a process called the FACE Method. The acronym stands for focus, ask, comment, and employ. Focus by locking in on the person's face and listening for their name. Then ask a question related to the person's name, like whether they prefer a nickname or which spelling they use. Then make some comment about the name while thinking about it in your head. If you met someone named Kristian, you might comment that you just read a book coauthored by someone with that name. Finally, use the name. Say something like "Nice to see you, Kristian."

There are a few other tactics you can use to cement that person's name in memory. One is simply to repeat it over and over again in your head. As you're interacting with a group keep thinking to yourself "John. John. John. John." It might sound silly, but it works. Another method is to translate the person's name into imagery. This works best for names like Ruby or Summer. In these cases, you could picture a big red gemstone or your favorite beach. You can also make more abstract associations for names, like noting that the Julia you just met has the same name as your graduate school roommate, or that Jennifer has the same name as your aunt. Associating the name with mental imagery is strengthened by the emotional connection you have with certain objects. It doesn't matter whether the association is negative or positive, only that it's there. One more technique is to employ mnemonics. Think "Graceful Gagan" or "Camacho loves nachos." The more you practice these techniques, the more they will become second nature.

Where Are You Headed?

If you're like the rest of us, you've gotten sidetracked while driving, taking the wrong turn because it's the one you usually take. Only when it's too late do you realize the mistake. These errors are the result of the brain's autopilot drawing from procedural memory. This type of memory is specialized to help you do all the repetitive tasks of your day while decreasing the load on attention, allowing you to attend to other, more novel things, like talking on the phone while driving or coming up with a new sentence while typing. Procedural memory is so efficient because it works unconsciously behind the scenes, freeing up consciousness for other tasks. When you learn new directions, you switch to using declarative memory. This type of memory is called declarative because it involves keeping track of information that you could describe to other people, such as the colors in a painting or the last ten presidents. So although much of our ambulatory life is spent in procedural memory mode, declarative memory plays a part. But how to internalize new directions given to you?

Your best bet is to use mental imagery. Many studies have shown it's possible to make mental models of a series of spoken directions. But it turns out there are a few ways to give directions. One way is to refer to streets connecting your various turning points, like "Go north on Chimney Rock Road until you get to Memorial Drive, then make a right. Continue for about four miles, then turn right on Shepard. Your next light is Kirby Drive. Make another right there. 1406 Kirby Drive is on your right." Another way is to refer to landmarks. For example, you could say something like "Keep going straight until you see a red Asian-styled house with stone walls in front, then turn right. Then go straight until you pass a large apartment complex. Turn right again. Take another right at the next light. Look for green gates on your right." People often disagree about which way is better. Some prefer hearing street names while others prefer the landmark approach. Still others like to use a combination of these methods.

It doesn't really matter which method you choose; your mental imagery will help you out. The key to remembering directions is to keep them as vivid as possible—vivid as in a strong mental image. The next time you listen for directions, try to visualize your turns while associating vivid images with each step. For example, rather than trying to remember street names or simple landmarks, try to build as much visual interest into the image as possible. "Keep going straight until you see an Asian house on your left . . ." Imagine you are taking a journey around the world, starting in America. You have to go east to your first destination, South Korea. You notice how different the homes look there. "Turn right. Go straight until you see a large apartment complex . . ." Your journey continues to China, where you stop in Hong Kong. Imagine a big city, so crowded that everyone is literally living on top of each other. "Then turn right on Shepard . . ." You decide to head to the countryside of Brazil where you see herds of white sheep, then you make a ninety-degree turn to the right to head back home. "Look for the green gates on your right." Now all you need to do is make it through customs. But since you don't have any bags with you, you get the green light to enter in no time.

It's not enough that you get to each landmark—you need to remember which way to go after that. So make sure to attach as much vividness to the direction you're supposed to turn as you do to each landmark.

Polyglots

Fostering meaningful connections between apparently arbitrary facts can help with other skills besides memory. The most successful language learning programs give you sufficient rudimentary skills to perform basic conversations in the new language within a week. They do that by using existing grammar and language composition skills and by fostering new memory associations. One example of this type of program is the Michel Thomas method. Michel Thomas was a polyglot linguist who developed and taught languages at a highly successful school in Beverly Hills called the Polyglot Institute. He claimed that his students could "achieve in three days what is not achieved in two to three years at any college." He later opened a second school in New York.

Thomas became known throughout the world after teaching college students with no previous experience the basics of French within five days on the BBC television science documentary *The Language Master.* His clients included celebrities such as Barbra Streisand, Woody Allen, and Grace Kelly.

The Michel Thomas program starts by making the students relax and not worry about practicing or consciously attempting to memorize anything they are taught. The teacher then introduces short phrases that represent the building blocks of the language, such as "es necesario," "no es aceptable," "no es posible para mí" (Spanish for "it is necessary," "it is not acceptable," and "it is not possible for me"). For the Romance and Germanic languages the method also makes use of commonalities between the Latin or Germanic influences on English and the target language when feasible. For example, in the Spanish program, English words ending in *-tion, -ible, -able,* and *-ary* are quickly transformed into their Spanish equivalents, for example, *situación, posible, aceptable,* and *necesario.* This gives the students a starting vocabulary of several thousand words and phrases.

The method also makes use of mnemonics. For example, in the Spanish program, the adjective *caro,* which means *expensive,* is remembered as "expensive car," the word *voy* (pronounced "boy"), which means *I go/am going,* is remembered as "oh boy oh boy oh boy," the term *tengo,* which means *I have,* is remembered as the Argentinian dance and style of music called tango, and the Spanish verb *estar,* which means *to be* (temporarily), is remembered as "to be a star." Once the backbone of a language has been mastered, proficiency can be achieved by memorizing the most commonly used words and by aiding comprehension by listening to native conversations in the language.

The more languages a student has already mastered, the faster a new language can be acquired. For example, students who already know Spanish can fairly easily learn the Romance languages Portuguese, Italian, French, and Romanian, which like Spanish descended from Vulgar Latin, and students who already know German can relatively quickly pick up the closely related Germanic languages Dutch, Danish, Swedish, Norwegian, and Icelandic, which like English descended from proto-Germanic. So, once one new language has been acquired, the speed with which a second or third or fourth language can be learned increases significantly.

The Michel Thomas method has been used to make commercial language-learning programs for French, German, Italian, Spanish, Russian, Arabic, Mandarin Chinese, Greek, Portuguese, Japanese, Dutch, and Polish. It may be a program of this kind that Daniel Tammet relied on when he learned Icelandic in one week on the television documentary *The Boy with the Incredible Brain*.

Designer Prodigies

The mnemonics presented above work well on the adult mind, but there are some specific types of mnemonics that work particularly well on young minds. As we have discussed, young brains are more plastic—they have the propensity to form new connections much more readily and rapidly—than adult ones.

Child therapist Peg Schwartz has developed a method to help children memorize passages of text that otherwise would be difficult to recall. When teaching a child to memorize new text, simply insert plausible errors on purpose. The method capitalizes on the connection between emotions and memory. The recognition that their teacher is making egregious errors gets them all fired up, enough that they begin to remember better. For example, consider the pledge of allegiance: "I pledge allegiance to the Flag of the United States of America, and to the Republic for which it stands, one Nation under God, indivisible, with liberty and justice for all." Peg would start, "I pledge allegiance . . . to my fairy," and the child would reply "Noooo, to the flag!"

Peg: ". . . of the United Arab Emirates . . ."
Child: "Noooo, of the United States of America!"
Peg: ". . . and to the rabbits . . ."
Child: "Nooo, the Republic!"
Peg: ". . . for which it sits down . . ."
Child: "Nooo, for which it stands!"
Peg: ". . . one nose . . ."
Child: "Noooo, one Nation!"
Peg: ". . . under grandma . . ."
Child: "Nooo, under God!"
Peg: ". . . invisible . . ."
Child: "Nooo, indivisible!"
Peg: ". . . with lizards . . ."
Child: "Nooo, with liberty!"
Peg: ". . . and justice for girls."
Child: "Nooo, for all!"

The idea is to insert words that sound similar to the correct words to remind the child of the real word. That creates a natural mnemonic association in the child's brain between the erroneous word and the real one. With some practice, the child will be able to recall the whole pledge without any priming.

But what if your child is too young to converse with you? One way of improving your child's mental abilities even before they learn to speak is to teach them baby sign language. It might sound crazy, but children can learn to sign as early as six months after birth. Baby sign language uses the same symbols as adult American Sign Language. The only difference between learning standard sign language and baby sign language is the pace at which the child is able to learn the symbols. Just be patient. It might take as long as two months for your child to start signing back to you. But that's with only five minutes of training per day.

Start out by using a few basic symbols that refer to objects or actions that occur daily. For instance, you might start by teaching your child the symbols for *mom, dad, milk,* and *eat.* To sign *mom,* simply place your right thumb against your chin while all your fingers are open. Sign *dad* by making the same motion, but tapping your forehead instead of your chin. *Milk* can be signed by squeezing and opening your right fist as though you're milking a cow. And *eat* is symbolized by putting your fingers together like you're stuffing a piece of food in your mouth, then tapping your lips.

Baby sign language has all sorts of benefits. Children who learn to communicate earlier are far less likely to suffer from emotional frustration. Rather than having to resort to temper tantrums to get what they want, babies can simply sign for it. More important, studies have shown that not only do infants who learn sign language communicate better with their parents, families who sign from infancy share stronger emotional bonds than those who don't. Baby sign language has also been associated with improved verbal communication skills during later years. One study compared a group of eleven-month-old babies who were taught sign language and received verbal training with another group of infants who received verbal training alone. Surprisingly, the more advanced talkers came from the signing group. At two years old, the signing group was still ahead by an average of three months. Research also suggests that learning to sign from a young age has positive benefits on general intelligence. A later study revisited the two groups from the previous study when they reached eight years of age. While the nonsigning group was of average intelligence, the signing group showed higher IQs, to the tune of twelve points on average. That's a tremendous difference, placing the signing children in the top 25 percent of intelligence for their age group! Baby sign language may also be beneficial for autistic children who aren't able to communicate verbally. So if you want your children to start ahead, teach them to sign.

Exceeding the Limit

We've been talking about algorithms that make us smarter, faster, and more rounded, more like the real Rain Man Kim Peek, who could remember most of what he had ever read. Armed with your new skills, consider enrolling in a memory competition. Think you're ready to participate in the World Memory Championship? Here are a few ways you could win: Memorize over 124 words in five minutes, 4,140 digits in thirty minutes, or twenty-eight decks of cards in an hour.

But how far can we go before we reach the limits of training? As we saw before, Jason Padgett receives information about the elaborate mathematical patterns he is seeing and drawing through vision, but he is not aware of the calculations his brain makes before producing the geometrical shapes. He turns his eyes toward something and receives the output in the form of a complex mathematical shape. As Jason puts it, "My conscious brain is a receiver of the result of the calculation made by my unconscious brain." Even Mark Nissen, who denies being born with any extraordinary mental abilities, swiftly uses imagery to access number storage space far outside his conscious awareness.

There are, of course, physical and biological limitations to our brain's abilities to calculate—the limitations to our conscious abilities and limitations imposed by dominant brain regions. We cannot consciously make hugely complex calculations in our heads or perform calculations that spit out a fractal in our field of vision. But we can use certain tricks to circumvent the restrictions that consciousness imposes. Using the algorithms provided above is the first step to reaching the mind's full potential.

CHAPTER 5

Smart Cookies

Human Calculators and Out-of-the-Blue Mathematicians

At the tender age of three Kim Ung-yong studied physics at Hanyang University and solved complicated differential equations. At the age of four he could read in Japanese, Korean, German, and English. He later learned Chinese, Spanish, Vietnamese, and Tagalog. His score on an IQ test designed for seven-year-olds at the time was 200. Now in his early fifties, Kim is recognized by the Guinness Book of World Records as having an IQ of 210, the highest ever measured.

Kim was a child with prodigious abilities and obsessions, and there are others like him. Sho Yano, who began college at age nine and entered medical school at age twelve, graduated in June 2012, at age twenty-one, from the Pritzker School of Medicine at the University of Chicago with an MD and a PhD in molecular genetics and cell biology. He then started his residency as a pediatric neurologist at the University of Chicago.

Graduating from medical school at age twenty-one might seem impressive, but Sho is not even the youngest person to do so. Balamurali

Ambati graduated from New York University at the age of thirteen, two years after graduating from high school and coauthoring his first book. Four years later, at age seventeen, he graduated from Mount Sinai School of Medicine. He completed an ophthalmology residency at Harvard University and later joined the medical faculty at the Medical College of Georgia.

Another absolutely amazing whiz kid, Arden Hayes, became world famous when he appeared on the television show *Jimmy Kimmel Live* on July 2, 2013, at age five. He was questioned about his knowledge of the US Presidents and recited Lincoln's Gettysburg Address. His interest in American history started at age two when he realized that he shared his birthday, January 30, with Franklin Roosevelt.

On November 5, 2013, Arden, still five, once again appeared on *Jimmy Kimmel Live*. This time he was quizzed on geography. The first thing he said when he was introduced to the map that would be used in the quizzing was that it didn't have South Sudan on it. The host started off with an easy question. How many countries are there? "one hundred ninety-six," Arden quickly replied. He was then shown the shape of a country and asked to identify the country and name its capital. "Paraguay," he said, and added that its capital is Asunción. Next he was asked to identify Yemen by its shape, and he did. He was then given a capital and asked to name the country. The capital was Riga. "Latvia," Arden announced. He was then once again shown a shape and asked what country it was. The young boy did not hesitate before pointing out that what he was shown was not a country, although it used to be Yugoslavia, and he then named the seven countries Yugoslavia was split up into. The kid was not even in kindergarten at the time. Amazing!

However, you don't need to be born a prodigy to achieve the abilities of prodigious children. Every child has a supermind. At least that's what Soundari Raj has set out to prove. Her math-training program has made it possible for the average child to calculate two hundred multi-digit sums in less than eight minutes—that's faster than 2.5 seconds per calculation. The Universal Concept of Mental Arithmetic Systems (UCMAS), the ones she and her colleagues use to train children's memory and thinking speed, use visual flash cards as well as the abacus, an ancient device typically made of wood and a bunch of free-sliding beads. A quick sketch of how to use the abacus to learn speed calculation can be found in the appendix.

For the first six months of the UCMAS program the child is taught how to use an abacus. Then flash cards are used to memorize how the different arrangements of beads correspond with different numbers. The goal is to be able to visualize how different calculations look on the abacus. By the end of the program the children have 450,000 image-number pairs stored in their memory. Some make use of their fingers to manipulate an imagined air abacus. Once they perfect their ability to recall these images and convert them into numbers, they add, subtract, and multiply large numbers with the help of the mental abacus. The UCMAS method thus turns children into little "rain men" by teaching them to use both sides of their brain: the right brain for the visuals and the abacus and the left side for the numbers. This sort of speed calculation is the basis for Flash Anzan, a contest in which individuals compete to add up fifteen three-digit numbers as fast as possible. The record? 1.7 seconds. The holder is a thirty-something school clerk named Takeo Sasano.

Another algorithm that almost anyone can train their brain to use allows for the multiplication of numbers at lightning speed. The crisscross method, as it is called, makes it easier to perform complex calculations in our heads by reducing multiplication to a series of simple steps. For example, say you want to multiply a couple of two-digit numbers. Depending on the digits in that number, the traditional pen and paper method of multiplication could be mentally exhausting.

$$7 \quad 9$$
$$\times 8 \quad 2$$

Step 1 $\vdots \; \vdots$ $2 \times 9 = 18$

$^{1}\underline{8}$

Step 2 $\times$ $(9 \times 8) + (7 \times 2) = 86^{+1}$

$^{8}\underline{7} 8$

Step 3 $\vdots \; \vdots$ $7 \times 8 = 56^{+8}$

$\underline{6 4} 7 8$

Result $\boxed{6,478}$

Various algorithms, such as the crisscross method, have been derived to allow for lightning-fast complex multiplication.

Using the method, two two-digit numbers are multiplied using a series of simple steps. To use the crisscross method, start by lining the two up vertically. Step 1: Multiply the two numbers in the last column. The ones column of the result is the last number in your answer. Save the number in the tens column of the result to carry over. Step 2: Multiply the top number in the right column with the bottom number in the left column and the top number in the left column with the bottom number in the right column. Add these two numbers together. Take the number in the ones column of the result and add it to the carryover from the first step. Place the number in the ones column of the resultant number to the left of the first number in your answer and save the number in the tens column to carry. Step 3: Multiply the numbers in the left column and add any carryover. The resultant number is the first two numbers in your answer.

The method takes some practice to master, but it makes mental multiplication more realistic for the average person, especially when using larger numbers. The method does not require particularly good memory skills. Almost any person can learn the 64 facts of single-digit multiplication by heart (not including $0 \times n$ and $1 \times n$). With an intact working memory and simple addition skills added to the mix, the crisscross method becomes very straightforward.

Before starting, you might find the method difficult and intimidating, because it still requires you to be able to keep track of various numbers and carry values. However, there are ways to practice this. One is to visualize the numbers in your mind's eye and work it out on a mental sheet of paper, to take advantage of your right hemisphere, as with the abacus. Within a few days of practice, two-digit numbers can be multiplied in a couple of seconds. More practice allows for even greater numbers to be multiplied with ease. At the beginning you will just get very fast at performing the steps of the crisscross method, but eventually it will become ingrained enough that you don't really go through each step, the way we don't sound out each letter when we read a word.

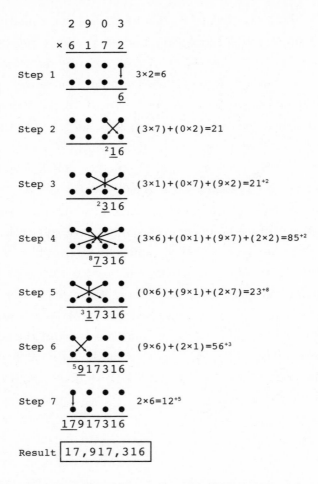

The crisscross method can be used to multiply any two numbers with great speed. In this example, two four-digit numbers are multiplied.

In *The Boy with the Incredible Brain*, Daniel Tammet calculates 37 to the power of 4 in a few seconds. With algorithms like the crisscross method, ordinary folks can accomplish the same feat very quickly. We begin with 37 squared. 7×7 gives us 49 and hence the last digit. $(3 \times 7) + (3 \times 7)$ equals 42. Adding 4 yields 46 and hence the second to the last digit. 3 squared plus 4 equals 13—the two remaining digits. We then need to calculate 1,369 squared using the same method, which amounts to 1,874,161.

Here is another way to do a limited type of quick multiplication. Multiply two two-digit numbers in your head within five minutes of practice (up to 20 × 20). Example: 15 × 14. Add the larger number and the second digit of the smaller number (15 + 4). That's 19. Multiply by 10 = 190. Multiply the two second digits of the original problem: 5 × 4 = 20. Add 190 and 20 to get 210.

Calendar Calculation

Recall the amazing cases of Orlando Serrell and Kim Peek. Both of these amazing savants were able to perform calendar calculations. For example, if you had asked about January 15, 2004, they would immediately have said "Thursday." This remarkable feat can be mastered easily with an algorithm. By learning an algorithm we are slowly teaching our brains to perform calendar calculations the same way Serrell and Peek's brain did it. At the beginning of a practice period, the algorithm will be like a recipe you consciously walk through, but with practice the brain will internalize it, and you won't need to consciously perform each step.

Here is one simple algorithm for performing calendar calculation, which is a very practical savant skill to have, especially when you practice it for the current year! Memorize one date for *a* Saturday in each month of 2015. That's a sequence of twelve numbers, for example:

2015

Jan—10		Jul—11
Feb—14		Aug—8
Mar—14		Sep—5
Apr—4		Oct—10
May—9		Nov—7
Jun—6		Dec—12

These are the so-called anchor dates. In your imagination you then assign the numbers 1 to 12 to each of your fingers and your two elbows. (If you merely want to use it as a party trick, you can cheat by actually writing a number on each of your finger tips and two on two of them.) When people say, for example, May 30, 2015, you go to finger 5 (May) and retrieve the number 9 from the table above. So, you know that May 9 (5/9) was a Saturday. The date you were asked about was May 30. 30 minus 2 is 28. 28 days amount to 4 weeks. So, since May 2 was a Saturday, so was May 30. You can quickly announce that May 30, 2015 was a Saturday.

Or someone says January 15, 2015 and you go to finger 1 (for January) and retrieve 10 from the table above. You know that January 10 was a Saturday. The 15th was five days later. So, January 15 was a Thursday.

Notice that the twelve anchor dates crucial to this algorithm aren't completely arbitrary. The majority of the number associations in the table in the appendix are rather easy to remember: for example 1/10, 2/14 (Valentine's Day), 3/14 (same as February), 4/4, 5/9 (standard 9–5 work hours reversed), 6/6, 7/11 (7-11 store), 8/8, 9/5 (standard work hours), 10/10, 11/7 (7-11 store reversed), and 12/12. There may be anchor dates that are even easier for you to remember, such as birthdays, anniversaries, or other special days in your personal life that can aid memory. Also, don't forget about other easy-to-remember dates such as April 15 (US tax day), July 4 (Independence Day), and December 25 (Christmas).

Of course, learning that, say, Thursday is five days later than Saturday is easier said than done. However, this part of the calculation gets easier with practice, as we become more familiar with thinking about the distances between the days. You will eventually get to a point at which you don't need to consciously count things up and add them together. Recall the case of the mathematician John Conway, a neurotypical person who has internalized these kinds of algorithms. Before

Conway gets down to work in the morning, he has to mentally calculate ten random calendar dates that the computer gives him. He calculates all ten dates in less than twenty seconds. Since he is averaging two seconds per calendar calculation, it is clear that Conway is not consciously going through each and every step of the algorithm. He initiates the main steps, and his brain unconsciously completes the work. The Doomsday method Conway developed and used to learn calendar calculation can be found in the appendix.

You can use the Doomsday method for dates of any year—you just need to remember which day of the week your anchor dates fall on for that year.

Using your ten fingers as a natural number system for the months also helps you internalize this algorithm more quickly because many of us already associate a number with each of the fingers. You are then relying not only on the mathematical calculations of the left hemisphere but also on the visualization abilities of the right hemisphere.

With practice, calendar calculation can become as automatic as bicycling or playing the piano. When you become experienced you don't need to think about riding your bicycle. You can ride it and think about something else. The processes underlying a skill change systematically with practice from deliberate, conscious activity to automatic activity. This also holds true for mental calculation. The initial, consciously controlled processes are more dependent on frontal lobe function than are automatic processes. Automatized arithmetic processing, on the other hand, requires only parietal neural processing. So, what the full internalization of an algorithm does is to switch the brain regions that process the steps in an algorithm from frontal and parietal regions of the brain to less high-level parietal regions of the brain. It thus makes the neural processing more similar to that performed by the brain of the autistic savant. Mental arithmetic is processed primarily in the left hemisphere in neurotypical individuals. However, this is consistent with the left-hemisphere defects that are typical in autistic savants. Most individuals with autism have damages to the left temporal lobe and the left frontal lobe but not necessarily to the left parietal lobe.

CHAPTER 6

Draw Like a Child

How to Turn Off the Nonartistic Brain

John Carter was a fifty-two-year old investment broker with a two-stroke golf handicap when he decided to give up his successful career to become a painter. John had no previous interest in art and had no prior knowledge of it. He would do whatever it took to avoid going to a museum or art exhibit. But suddenly he felt extremely drawn to the visual arts. He felt so drawn that he had to create. He reported that he was overtaken by visions that needed to be captured on canvas.

His friends at first thought that the change in his life and personality was a part of his prolonged grieving after his wife's death. But they realized he was serious about painting when he actually quit his job and moved into an old loft and began painting full time. He also changed his style of clothing from formal business attire to colorful outfits. He would often wear purple shirts and yellow pants and use the same colors in his paintings.

After he spent a few months making what seemed like rather meaningless strokes on a canvas, his abstract paintings began to take form. The paintings radiated their own special beauty. John was soon winning various art awards and had his work displayed in a gallery in New York City. At that point John could no longer remember the meaning of many common words. As his language abilities deteriorated, his paintings became better and better. Eventually he lost all ability to speak and name objects he could draw, but his drawing skills kept improving. Sadly, his dementia eventually progressed to the point that he lost the ability to draw as well, and he died at age sixty-eight.

Losing Your Mind

John suffered from the fatal condition frontotemporal dementia. As we saw in chapter 1, the condition affects the temporal and frontal regions of the left hemisphere, the areas associated with spoken language and arithmetic. Individuals with this condition sometimes develop remarkable artistic talent due to weakening executive areas of the brain that hinder hyperactivity in neural regions associated with extraordinary abilities.

Frontotemporal dementia differs from the more commonly known form of dementia known as Alzheimer's disease. Alzheimer's disease is a form of dementia that arises when certain proteins, called beta-amyloids, clump together to form plaques between nerve cells in the hippocampus, the brain's main center for storage-based short-term memory. At the same time, twisted strands of another protein, called tau, begin to form inside the neurons, which prevents them from getting nutrients. Eventually, the hippocampus will wither and die. A smaller and clotted hippocampus not only hinders storage of information for the short term, it prevents new information from moving into the cerebral cortex, an outer brain region that normally stores information for the long term. Frontotemporal dementia does not by itself affect the hippocampus. Instead, it affects the system that executes top-down control of lower-level neural processing. Once the top-down control system is disabled, the lower-level neural processing becomes more active than usual. It is this additional activity in lower brain regions that gives rise to the remarkable creative abilities seen in people afflicted with this condition.

Neurologist Bruce Miller has reported five cases of individuals with acquired savant syndrome as a result of frontotemporal dementia. The condition typically gives rise to artistic abilities that are specific to the visual domain, never manifesting in the verbal sphere. As they slowly lose their language function, patients with frontotemporal dementia often develop extraordinary new artistic abilities. There are certain similarities among most patients with this condition. They tend to create artwork that is realistic in style, only rarely abstract. The images typically depict scenes from the past, suggesting that while they might be unable to speak about the past, their visual memory is intact at least for a while. Most of the patients who begin to draw or paint after the onset of dementia have no prior interest in art.

The type of frontotemporal dementia that tends to give rise to savant skills is also known as semantic dementia. It's a degeneration of language areas and areas that process the concept of what words are, particularly the left anterior temporal lobe. Although this disease may hit both sides of the brain, the left side that is responsible for language is much more severely affected. The degeneration normally also deprives the patients of the ability to recognize faces.

The parietal lobe on top of the head, however, remains intact, as do large portions of the frontal lobes, so most of these patients can imitate with incredible precision in drawing or painting. They also tend to develop a compulsive behavior that inspires them to express themselves artistically and continue to do so over and over again, up to twenty to thirty times. They simply keep drawing until they are satisfied with what they see.

According to Miller, these cases indicate that the most important parts of visual art lie in the right hemisphere and include three key abilities, namely visual constructive ability, spatial attention, and internal representation. The left hemisphere is much more involved in artistic forms such as symbolism and abstract representation. As people with frontotemporal dementia lose the ability to speak and to process linguistic meaning of things, there's a loss of higher-order processing that goes on in the anterior temporal lobe, as is also seen in autistic savants.

The Brain Arrangement That Unlocks Extraordinary Ability

The case of people with frontotemporal dementia gives some insight into the particular arrangement that unlocks extraordinary creativity, especially in the visual arts. In neurotypical people, high-level brain regions that process big-picture features, particularly the prefrontal lobe, dominate over the motor-sensory regions that process literal features. When you look at a landscape, you don't see a collection of literal details such as gradients of texture, shading, and immediate size—you see buildings and trees casting shadows, roads that look smaller and smoother as they stretch toward the horizon, and tall mountains in the distance that appear pale and hazy because they are so far away. It's literal features such as texture and shading that make it possible for us to see three-dimensional objects against a background rather than just two-dimensional patterns embedded in a sea of colors. Your brain uses these cues to assemble a three-dimensional interpretation of the world out of countless individual visual details.

But if you were to try to paint this landscape, the importance of these visual cues becomes clear—the more accurately you can replicate them, the more realistic and three dimensional your painting will appear. But it's not so easy to isolate these literal details after our higher-level brain has already interpreted them as information about distance or shadow.

Look at the left-hand image below, of a cylindrical shape casting a shadow across a checkerboard. If you were to reproduce it in a painting or drawing, would you use a dark color for square A and a lighter one for square B? Most of us who are not trained artists would, but as you can see from the image on the right, the squares are in fact the same color. Our brains automatically process the contextual information (the checkerboard, the shadow), and so we perceive square B as being a lighter color.

Edward Adelson's Checkerboard Illusion. The visual system adjusts for the apparent differences in the spectral power distribution of the illuminant, which leads us to perceive A and B as differently colored.

But individuals with frontotemporal degeneration do not have the same sort of top-down inhibition of access to raw literal information—that the squares are in fact the same color—and can therefore replicate it more accurately. The brain is filtering out literal information in order to generate a bigger picture.

Creative and artistic savant skills as well as extreme recall abilities are mostly nonverbal skills, that is, skills that help access literal detail.

They rely on accessing very literal details of a scene or on remembering what can seem to be rather minute details of the incoming information rather than the big picture that most of us remember. For example, when watching a documentary about a famous person, most of us will remember the big-picture narrative, the person's life story, but not the dates of important events or the color of her clothing in different scenes. This is because we tend to interpret the information presented to us, and remember only the outcome of the interpretation. Once a brain defect makes the interpretation of literal information more difficult, it is easier for the brain to access and store the details of the incoming information. So, brain defects that cause increased access to literal uninterpreted information can dramatically increase creativity and artistic acumen as well as the ability to remember detailed information. While the brain takes in the same amount of information as the neurotypical brain, the key difference is that in the case of frontotemporal dementia the "picture filter" is broken, so most literal detail makes it into the individual's conscious awareness.

There are several ways that brain defects can hinder big-picture interpretation of raw sensory information. When long-range brain connections, including information flow between the two hemispheres, is blocked or reduced, a big-picture interpretation becomes more difficult, and conscious access to literal sensory information increases. Kim Peek is a good example of a person with these sorts of brain abnormalities. A similar thing happens when executive regions, such as frontotemporal regions in the left hemisphere, are damaged or inhibited. In both cases, the brain returns to a preconceptual, child-like, autistic-like, literal state. A child's brain is preconceptual and right-hemisphere dominated. Young children see things more literally than adults and do not apply abstract concepts and meaningful labels to things. With maturity, neurotypical individuals develop left-hemisphere skills, such as language, and apply top-down processing, effectively generating a simplified, schematic model of reality.

The most extreme example of a brain lacking top-down control is perhaps the case of Michelle Mack, a woman who was born with only the right hemisphere of her brain, as detailed by Norman Doidge in his book *The Brain That Changes Itself.* Over time Michelle's right hemisphere did develop many left-hemisphere skills. She learned to speak, had a part-time job and a family. But there were also many skills that were never fully developed. Her right visual field was limited, so she had a hard time seeing things coming from her right. Her capacity to abstract and apply semantic labels to objects was also severely restricted. For example, a young child can easily abstract away from the individual differences between particular bananas and learn to apply the label *banana* whenever he or she encounters a banana. Michelle had difficulties with this type of abstraction process.

Furthermore, like autistic individuals, she developed sensory hypersensitivity. She had an excellent memory for literal details and had savant-level skills in calendar calculation. For example, she could tell by memory what day of the week a date was within the previous eighteen years. She had to use an algorithm to calculate dates earlier than that, but she could nonetheless still do it quickly and accurately.

An interesting question arises about whether we can undo the inhibition of access to low-level raw information in normal individuals. The answer is yes, at least to some extent. The neurotypical brain possesses all the necessary information required to draw, play music, or retain information for long time periods, but most people are unable to access it for creative purposes. We are hindered by the "big picture," the concepts and meaningful labels and prior connections that we have acquired through learning. However, with training, we can learn to turn off (or at least turn down) the activity of our meaning-attaching, interpretive left hemisphere and gain access to lower-level information in the brain.

Learning to draw or paint realistically involves learning how to block the way our brains automatically move from the two-dimensional image of the external world that is imprinted on the retina to a finished three-dimensional image. When you are driving down a road with equal-sized trees on each side, it is important that you see the trees realistically, that is, as equal-sized rather than as taking up different proportions of the visual field, but this is not good if you want to paint the landscape. The light from the equal-sized trees does not make equal-sized marks on your retina. The light from the trees that are closer leaves much larger imprints than that from trees that are farther away.

The brain's visual system and the left temporal lobe automatically translate the different-sized bits of light on your retina into a perception of equal-sized trees, with some nearer and some farther away. This is convenient when moving about in our external environment. It allows us to drive without hitting the trees, quickly estimating how far we have to go before making the next turn, or—more important for our evolutionary ancestors—how quickly we could escape up into a tree if we needed to avoid danger. It also prevents us from being overwhelmed by sensory details, which as we have seen can lead to a sensory overload and a need for rigid routines to ensure that there is some structure to reality.

But a good drawer or painter of realism must be able to see trees the way they are actually imprinted on the retina. One way to help yourself begin to see the literal, individual details of what you want to draw is to obscure the meaning of the overall picture in some way. If you are trying to draw a copy of a photo, for instance, you could turn it upside down to distance yourself from the immediate meaning of the photo. Right side up, it might be a photo of a horse, but upside down, it is a collection of lines and shapes that don't correspond to anything you are used to seeing. It requires a little bit of practice to obscure the meaning by looking at something but once people experience it, they usually find it rather simple and even addictive.

There are lots of other techniques you can use to distract the left hemisphere of your brain, which constantly wants to attach meaning to everything it sees. With practice anyone can learn to draw well within six weeks. In addition to viewing your subject upside down, you could try looking at it in a mirror. Or you could draw the negative space around your subject. For example, instead of drawing a pair of scissors, you draw the space surrounding the pair of scissors. This also takes your attention away from the meaning associated with the object you are drawing. To get the hang of perspective in drawing, you can use a handy measure of comparison: your own fingers. For example, if you are looking at people sitting at a table, they all seem to look more or less the same size. But if you measure them in front your eyes with your thumb and index finger as if you were picking them up, you will get a good sense of how small the more distant people to be drawn are. The "two balls illusion" nicely illustrates size constancy versus relative size:

The Two Balls Illusion. The ball depicted as farther away looks much larger than the ball in front of us, even though they make the same imprint on the retina. (Reprinted by permission from Macmillan Publishers Ltd: NATURE NEUROSCIENCE, "The representation of perceived angular size in human primary visual cortex," Scott O. Murray, Huseyin Boyacti, and Daniel Kersten, copyright 2006.)

The reason the illusion works is that if we had been in a natural environment rather than looking at a picture, the top ball would have been much larger than the bottom ball. So, the brain automatically adjusts for what it regards as a difference in distance between the two objects. It takes effort to see the two balls as having the same size, because seeing them as same-sized requires attending to literal features that are normally overridden by our dominant left hemisphere, which automatically interprets raw data and assigns meaning to them.

In general the trick to good drawing is to practice a visual technique that researcher Betty Edwards calls the "picture plane." It's an imaginary, transparent, framed window that is sitting in front of your face and through which you look. The picture plane shows everything in an egocentric way—that is, a way that relates to your own perspective. The equal-sized trees along the road become trees taking up different proportions of the visual field. A coin displayed to you sideways has an elliptical shape placed in a plane perpendicular to the line of your sight. So, the coin is represented as being elliptical from *here*. A uniformly white wall becomes a white wall with gray patches, where the gray patches are the shading. Whereas the brain uses shading to calculate three-dimensional objects, the artist accesses shading and other literal details directly and uses them to portray volume and dimensionality. The picture plane is attached to your eyes. So, if you turn, so does the picture plane.

Two Ways of Seeing

When you practice looking through the picture plane, you use your dorsal stream, the visual system that allows you to act on what you see. The dorsal stream kicks in automatically when you do something like reach out and grasp an object. It's distinct from the ventral stream, which is what allows you to perceive the world. When you see a cat across the street, it's your ventral stream that allows you to identify it as a cat. Both streams start in the primary visual cortex on the back of the head. The ventral "perceptual" stream runs into the temporal lobe on the side of the head and then connects to other temporal and frontal lobe structures that are responsible for short-term memory, decision-making, and rational thinking. The dorsal "action" stream runs upward through the visual areas into the parietal lobe on top of the head and continues until it makes contact with the primary somatosensory cortex and the primary motor cortex, responsible for planning and executing physical action. But there are also several pathways to the dorsal stream directly from the eye, for example, through subcortical brain regions known as the interlaminar, the superior culliculus, and the pulvinar. The subcortical pathways allow the information to enter the dorsal stream very quickly.

The dorsal "action" stream normally operates in the absence of visual awareness. For example, you don't look at your pants and shirt in the morning and then consciously decide how you need to move your hands and fingers in order to put them on. The movements of your hands when you act quickly are automatic and do not require any conscious input.

The dorsal stream is an older visual system, evolutionarily speaking, than the system that allows us to consciously see the world. Vision did not initially evolve as a system that allowed organisms to see the world as humans do today. It began as a way of coordinating physical movement. Many nonmammalian animals have rather complicated visual systems for performing actions. Frogs, for example, have at least five different visual systems that perform different functions associated with motion and action. One reason the ventral "perception" stream may have evolved in organisms that already have the ability to reach out and grab things is that this action requires the ability to perceive the constant, absolute size of an object in order for the creature to fold the part of their body that does the grasping or catching correctly. For example, if you want to grasp a stick, your hand needs to be folded one way, and if you want to grasp an apple, it needs to be folded a different way. A frog has to recognize how to fold its tongue around the fly it wants to eat. But this requires perceiving the object in a way that abstracts away from your present subject-dependent point of view.

The ventral "perception" stream processes information about color, shape, size, and relations between an object and its surroundings in a way that transcends different viewpoints. Unlike the dorsal "action" stream, it thus constructs a representation of the world that is constant and independent of viewpoint. But a representation that is constant and independent of perspective does not suffice for picking up a glass of beer or operating a computer mouse. For that purpose we need the dorsal "action" stream, which combines information about absolute size and viewpoint-dependent properties, such as the relative orientation of the object.

Evidence for the two visual streams comes from studies indicating that lesions to the dorsal stream can impair visuomotor control while leaving visual perception intact, whereas lesions to the ventral stream can impair visual perception while leaving visuomotor control intact. One of Milner and Goodale's patients, D.F., who had damage to the ventral stream after carbon-monoxide poisoning, could consciously see some color and texture of an object placed in front of her, but was unable to identify the object or recognize its shape. However, her dorsal "action" stream was intact. She couldn't describe a slot in the wall, but she could easily insert a card into it. And she could easily pick up a rectangular block, despite being completely unable to describe what she was doing.

Milner and Goodale argued that because D.F. had no visual awareness of the shape or identity of objects, she could not have been relying on working memory when she reached out to grasp them. The information in the dorsal "action" stream, they concluded, is not stored in working memory.

Milner and Goodale also looked at optic ataxia patients. Optic ataxia results from damage to the dorsal action stream. Patients with this condition are unable to automatically adjust their handgrip to the size of objects they try to grasp. Milner and Goodale's optic ataxic patient I.G., for example, had difficulty adjusting the shape of his hand to an object's size and shape when the task was to grab the object as soon as he saw it. In order to grab objects, patients with optic ataxia reach to an object with a completely flat hand and only close their hand around the object when they feel it.

The features that we need to recognize in order to quickly reach toward and grasp an object are also the features we need to learn to see and possibly remember in order to draw well. But the egocentric representations computed by the dorsal action stream are highly transient and are not normally stored in memory. This is why we cannot recall all the details we see when moving quickly through a scene.

Savants with extreme artistic abilities differ from most of us in this regard. Autistic savant Stephen Wiltshire, for example, drew an extremely accurate sketch of a four-square-mile section of London, including twelve major landmarks and two hundred other buildings, after a twelve-minute helicopter ride through the area. Likewise, a three-year-old autistic girl named Nadia became famous for her ability to sketch spectacularly detailed galloping horses and riders from memory. One day, she suddenly picked up a pen and started to draw. With no prior training, she created sketches of horses that only a trained artist could match. But her approach was different from that of many trained artists who begin with an outline. She began with random details. A hoof, the horse's mane, its harness. Finally she would draw lines connecting these free-floating features.

Both visual systems—the ventral and the dorsal streams—are primarily right-hemisphere-based. However, to acquire some of the skills of the amazing autistic savants who can draw or sculpt with astonishing accuracy, we must find a way to access the literal details of objects. We need to switch from a ventral-stream way of seeing to a dorsal-stream way of seeing. This is just what nonsavants who can draw or sculpt very skillfully do. Switching from a ventral-stream way of seeing to a dorsal-stream way of seeing takes a bit of practice, but nowhere near the ten thousand hours that would be required if we were simply practicing without a strategy.

We have now gotten a glimpse into what the magic arrangement of the brain is that unlocks extraordinary abilities such as the savant skills. Interestingly enough, we are all in a condition that is like frontotemporal dementia—a more dorsal-stream state—when we sleep.

CHAPTER 7

Not a Worry in the World

Problem-Solving Strategies

In the summer of 1816, Mary Wollstonecraft Godwin, the author of the famous work *Frankenstein*, hit on the idea for the book in a dream. In the introduction to the book, she writes: "With shut eyes, but acute mental vision, I saw the pale student of unhallowed arts kneeling beside the thing he had put together. I saw the hideous phantasm of a man stretched out, and then, on the working of some powerful engine, show signs of life, and stir with an uneasy, half-vital motion."

In 1903, Otto Loewi had the idea that there might be a chemical transmission of nerve impulses, but he was unable to prove it until one night when he dreamed about the design of an experiment that was able to prove his theory. This work led to a Nobel Prize.

In 1964, golfer Jack Nicklaus had had a run of poor scores when he dreamed about his golf swing and discovered that he had been holding the club incorrectly. When he woke the next morning he was able to correct it and shot a 68.

It might come as a surprise that sleep can be a time of great accomplishment. After all, how could you get anything done reclined and paralyzed? But people are capable of strange things while they sleep. Some talk. Some kick their partners due to restless leg syndrome. Some snore and temporarily stop breathing. And then there are people who get out of their beds while still asleep and complete complicated actions. We call it sleepwalking, but walk is not all they do. One Australian woman would regularly get out of bed at night and have sex with strangers. Former chef Rob Wood cooked spaghetti Bolognese, omelets, and fish and chips during his sleep. The late Australian visionary, inventor, and artist Myra Juliet Farrell wrote the solutions to various problems in her sleep. Nurse and renowned artist Lee Hadwin produces unbelievable artworks while asleep. Computer expert Ian Armstrong was observed by his wife mowing the lawn naked at two a.m. while asleep. And it was recently discovered in a sleep lab that I regularly sleep-work on my laptop, which includes sending fairly coherent and meaningful e-mails to people. This is also known as zzz-mailing.

Sleepwalking has even been used as a defense for murder. The first case was Albert Tirrell. In 1846, he slit the throat of a Boston prostitute, set fire to a brothel, and took off to New Orleans. The defense was able to convince the jury that Tirrell was a chronic sleepwalker and could have committed the crimes while sleepwalking. Perhaps the most interesting and convincing case of a killer who was sleepwalking is that of Kenneth Parks, then a twenty-three-year-old father and husband from Toronto. One early morning in May 1987, Kenneth got out of bed and drove twenty-three kilometers from Pickering to the house of his wife's parents, where he proceeded to bludgeon them to death with a tire iron. After careful examination of the case, the experts could find no other explanation of the crime than sleepwalking. He was later acquitted of the double homicide in court.

Sleepwalkers typically have no memory of their sleepwalking

activities, as their consciousness is in a very low state of activity, during which it is nearly impossible to retrieve memories. Although the eyes of sleepwalkers are open, their gaze is diffuse. The sleepwalking normally occurs during deep, dreamless sleep.

Most of us sleep to get away from it all, but it turns out that sleep can be as productive as a day at the office. Many successful individuals have found a way to turn their dreams into office hours, allowing them to discover intellectual or artistic masterpieces despite the lack of any waking effort. Some of these individuals stand by idly, observing their imaginary life fly by with great intrigue. But others have taught themselves to command their own dreams, taking advantage of the fuzzy borders between conscious and unconscious thoughts that serve the dream world.

Sleeping In

Given that many of us spend about one-third of our lives in bed, it might come as a surprise that we still don't know why we, along with most other creatures in the animal kingdom, sleep. Even the most well-known sleep researcher, William Dement, who founded Stanford University's Sleep Research Center, understands how much progress there is to be made. "As far as I know, the only reason we need to sleep that is really, really solid is because we get sleepy," he says. Although we've yet to figure out why it exists, the science of what happens during sleep gives us a clue: The phenomenon appears to be the brain's way of organizing itself, leading to better connections among the information pathways we use most frequently.

We do know sleep is essential to our being. If it weren't, then we'd expect to find animals that don't need to sleep at all. But we haven't. Aside from a few very simple invertebrates lacking brains, all animals sleep. It was once thought that sleep is the body's way of conserving energy, for it's reasonable to think that we expend far less energy when we're dormant. There are a few reasons this explanation is inadequate. First off, the sleep state only reduces our metabolism by a small margin. We hardly save enough energy to justify losing one third of potential hunting and gathering time in the process. Second, even animals in hibernation, a period of complete dormancy, transition in and out of sleep states. This transition requires that they regulate their body temperatures appropriately. But this process of regulating body temperature uses more energy than the total sleep period saves, leading to a net loss of energy for each sleep period. So in some instances sleep actually uses more energy than the waking state.

There is some evidence that sleep might be the time at which the brain does its "custodial" work. Just like all other organ systems, the brain is constantly in flux. Underutilized synaptic connections break and new ones emerge. Waste products are shuffled along through the bloodstream to be eliminated by the bowels. It appears that sleep might be the time that all of this occurs. One recent study published in an October 2013 issue of the journal *Science* showed that although the brain clears metabolic waste products all the time, it does so much faster in the sleeping state than in the waking state. Sleep also affects the rate at which our bodies heal. Numerous studies have found that sleep deficits are connected to things like decreased white blood cell count. White blood cells are the body's first line of defense against foreign invaders like infectious diseases. Sleep deprivation is also associated with increased tumor growth rate and even decreased growth hormone secretion in men.

Sleep also appears to be crucial in the formation and retention of memories. In a study published in the November 2007 issue of the journal *Neuropsychology* it was found that sleep deprivation can temporarily impede working memory by as much as 38 percent. As we learned in chapter 4, working memory is crucial for keeping track of all the information we use to make conscious decisions. Sleep also appears to be the time when the brain works on storing newly learned information to long-term memory. Sleep after learning encourages the growth of dendritic protrusions. These molecules work like tiny wires that connect brain cells to facilitate the passage of information. So getting a good night's rest is required for both proper storage and retrieval of information from memory. Sleep is important for our superminds.

That said, how and how much we sleep varies radically from individual to individual. Leonardo da Vinci, Thomas Edison, Nikola Tesla, Buckminster Fuller, and Margaret Thatcher were all fine with getting as little as four hours of sleep per night. Some people sleep on a Biphasic Schedule, that is, they sleep three to four hours, then they wake for about one hour and then they sleep three to four hours again. Others sleep on a Dymaxion Sleep Schedule. They sleep four times a day for thirty minutes. Yet others sleep in accordance with the Uberman Schedule, which consists of taking six twenty-minute naps per day. Still others follow an Everyman Schedule, sleeping three hours during the night and then taking three twenty-minute naps during the day. Getting enough sleep depends on what kind of sleeper you are. But you do need to get enough.

According to recent research, lack of proper sleep can lead to symptoms that mimic some of the bizarre symptoms caused by brain damage, such as hemispatial neglect. An individual afflicted with hemispatial neglect loses awareness of one side of the body. For example, she might brush her teeth only on one side, dress one side of her body, or even deny that limbs on one side of her body are hers. Hemispatial neglect is often the result of a stroke, when blood is cut off to a portion of one of the brain's hemispheres. The lack of blood flow prevents brain cells from getting the oxygen crucial to their operation, causing them to die off. Evidently, some of these neurons are crucial for spatial awareness. Symptoms are usually exhibited on the contralateral side. For example, damage to the left hemisphere results in spatial neglect of the right-hand side, and vice versa. But not always. In rare cases it's possible that the neglect occurs on the same side as the lesioned area.

Individuals with hemispatial neglect can see perfectly well, they just don't recognize anything on one side of the visual field. The condition is likely what inspired the famous line in *Zoolander* when Ben Stiller's character explains his freezing on the runway. "I'm not an ambi-turner. I can't turn left," he yells in despair. Though it might sound ridiculous, my coauthor Kristian, in fact, suffers from spatial neglect when he has a severe migraine. Except he can't turn right. The feeling is akin to the tip-of-the-tongue phenomenon. Despite knowing what he's trying to do, he simply cannot recognize it's possible to do anything in a clockwise direction. This can be particularly frustrating when trying to do activities that require dexterity, like cutting a steak. To cut a V-shaped section, you first start by making a cut angled to the left. But if you can't turn right, you have to find another way to make the cut angled in the other direction. So you turn the plate counter-clockwise by a few degrees. But sometimes you overestimate how much turning is needed. So now you're stuck turning the plate all the way around so you can try again. Eating becomes painful. Fortunately, the symptoms finally fade as he transitions from the pre-migraine aura to the pain period. If the brain mimics brain damage in other ways, it could mean that what happens to our brains during sleep is similar to what happens when someone sustains a brain injury that leads to savant syndrome.

A form of hemispatial neglect appears to occur in all of us as we drift off to sleep. As we slowly transition from consciousness to the sleep state, our spatial attention shifts to the right, as evidenced by a recent study showing that people are more likely to misattribute sounds occurring on the left side of the body as occurring on the right. In other words, we start to have experiences much like Zoolander. Initially it might be scary to think our minds are reduced to the operations of damaged brains. Fortunately, these deficits are only temporary. All you have to do to get rid of them is wake up. But spending some time in the brain-damaged state isn't always so bad. After all, it's brought success to individuals like Jason Padgett and Derek Amato. Imagine dozing off and suddenly being able to understand complex mathematics or play the piano. The latest research indicates that this might indeed be possible. Neurologist Andrei Medvedev and colleagues reported at the 2012 meeting of the Society for Neuroscience that, during sleep, the left-hemisphere activity decreases significantly while right-hemisphere activity buzzes along. The right hemisphere is often thought to be the side of the brain responsible for creative thinking.

A myriad of other studies support the conjecture that creative right-brain processes are highly active during sleep. Individuals deprived of sleep for a total of thirty-two hours exhibited a marked decrease in performance on a creative thinking test. In particular, the participants were less flexible and less original with their strategies at solving different puzzles. Studies also indicate that sleep can facilitate the recognition of hidden patterns, especially when the individual had been working on the problem before taking a snooze.

It also appears that the sleeping brain might use different processes than the waking brain when processing the same information. Neurophysiologist Matthew P. Walker and colleagues looked at what happens when participants are asked to complete a creative cognitive task after being awoken from different stages of the sleep cycles. Participants first completed a computerized anagram task to figure out their baseline score. An anagram is a word that contains all of the letters that spell a different word. For example, the letters in *Doctor Who* can be rearranged to spell *Torchwood*, the name of a spin-off from the original show's 2005 revival. The test measured how many anagrams participants could solve in under ten seconds each. The study found that participants perform as well as in the waking state on the anagram test when they were woken from particular stages in the sleep cycle, either right after sleep or after entering deep sleep. Given that the sleep state and waking states definitely are distinct, this means that the brain can use different processes to perform the same tasks.

These discoveries explain the "a-ha!" moments many of us experience when we dream. The reduction of left-hemisphere activity in combination with increased right-hemisphere activity puts our creative right brain at an advantage, allowing us to recognize its insights. In chapter 9, we'll see how scientists are using technology to artificially induce this creative state.

Most of the ideas we get when we are asleep, of course, are useless. But occasionally they will be useful. There is a long list of scientists and artists who discovered the core of their theories while asleep. Perhaps the most famous case of an artist who created while asleep is Keith Richards of the Rolling Stones. In 1965 Keith passed out with his guitar and a tape recorder in his hotel bed. When he woke up, he saw that the tape had run to the end. He rewound it and pressed play. And there was the beginning of a song that would become very famous, followed by forty minutes of snoring. The opening Keith had recorded in his sleep was the opening to the tune "(I Can't Get No) Satisfaction."

Even if we don't gain a whole new skill set while asleep, it can be a tool for reaching superhuman potential. The key is dreaming. This most ubiquitous but least understood aspect of sleep can let you experience being a famous painter or musical virtuoso, if only for a few hours per night.

The Dream Narrative

Dreams often seem like very wacky narratives that string together people and events from waking life in a bizarre order. Impossible things can occur. In the 1970s Harvard psychiatrist J. Allan Hobson famously argued against Sigmund Freud's theory of dreams as symbols of hidden wishes and desires. Hobson thought that we experience dreams as wacky because our dream experiences are a "patch job." They are meaningless narratives patched together to make sense of basic automatic responses, such as biochemical changes and spontaneous electric pulses, coming from the brain stem. While our brains eventually create these narratives, the narratives don't mean anything, said Hobson. The narratives enter the scene only once biochemical and pulse changes have occurred in the deep parts of the brain, and we enter a state in which we can use language regions and logic regions of the brain to make a story out of the electrical signals and the chemicals.

The received view now is somewhere in between: Dreams are neither strings of random symbols nor narrative presentations of hidden unconscious desires and wishes, but brain narratives that more or less openly reveal wishes and concerns in real life. G. William Domhoff, a professor at the University of California, Santa Cruz, has shown that a typical dream report provides a coherent, clear, and detailed account of a realistic situation and the dreamer's emotional responses to the events that transpire in the dream. About 50 percent of dreams include specific emotions, and about 80 percent of the emotions that appear in dreams are negative emotions, such as embarrassment, sadness, anger, aggression, and fear.

It is the REM sleep cycle that holds the key to being able to solve problems or increase creativity while we're asleep, because dreams deal with our daily problems. Dreams typically emerge during the REM sleep cycle.

Sleep normally occurs in ninety-minute cycles, each of which has about eighty minutes of non-REM sleep and ten minutes of REM sleep. Non-REM sleep occurs in three phases. The first phase is the border between sleep and wakefulness. It's right about the point when you stop counting sheep. Your muscles are still active, but your eyes start falling closed. Most of your brain waves during this stage fall into what is called the "alpha range," which ranges from 8–15 Hz and occurs mostly in back of the brain on both sides, but with a higher amplitude in the right hemisphere (or the left, if the right side is dominant—occurring in 18.8 percent of left-handed individuals). These are the same brain waves observed during a coma.

Then you enter the second phase. During the second stage the majority of your brain regions fire in the theta range, which ranges from 4–7 Hz. In adults these waves are seen mainly during sleep, but they also occur in wakeful children and sleepy adults who are trying to resist doing something, for example, eating an additional cookie.

These brain waves are also observed when people become unconscious after a brain lesion and during so-called vegetative states. The signs of stage 2 non-REM sleep are a series of brain wave patterns called sleep spindles followed by K-complexes. Sleep spindles appear as little electrical oscillations on an EEG scan, while K-complexes look like big spikes. They are thought to serve the function of telling the brain to relax, allowing it to remain asleep, which prepares it for the next stage.

The third state of non-REM sleep is characterized by the relaxation of muscles and increased difficulty of waking up. In fact, the sleeper will successfully ignore many stimuli in the environment, including car horns and snoring lovers. In this phrase the brain normally fires at an extremely low frequency, which is reached when 20–50 percent of all brain activity is in the delta range. This is anything below 4 Hz and occurs mostly in the front of the brain in adults and in the back of the brain in children. It is also the main form of brain wave activity in infants and has also been observed during some continuous attention tasks. It is very common when people become unconscious after a brain lesion and during vegetative states. Some add a stage 4 to the first three stages, which is classified as 50 percent or higher delta activity.

REM sleep is at the tail end of the sleep cycle. Usually, the brain reenters stages 3 and 2 and possibly stage 1 before reaching REM sleep. The REM stage is also the state after which we normally wake up when we are done sleeping. REM stands for rapid eye movement, named thus for the rapid fluttering of the eyes observed during this stage. At this point the muscles become paralyzed. Brain activity elevates tremendously, causing it to use more oxygen than it does when it's awake.

During REM sleep the brain fires in the gamma and beta range, which ranges from 16–32+ Hz and signals consciousness. These kinds of low-amplitude beta waves also occur frontally during active thinking, deep concentration, anxiety, and high alert. Bursts of beta activity in the motor regions of the brain on top of the head are associated with movements. Gamma waves occur mostly in the somatosensory cortex involved in touch and are involved in cross-modal sensory processing, such as when the sound of a dog barking is combined with the sight of a dog barking. Gamma waves are also involved in some forms of short-term memory.

The series of non-REM stages followed by the REM period make up one sleep cycle. A typical night will see four or five of these cycles, although the length of time for each may vary. And the four phases of sleep with these particular characteristics are just the typical pattern— the cycles may vary quite considerably among individuals. For example, the sleep study I participated in revealed that my brain often does not exhibit decreased levels of activity when I am completely asleep. I also have rapid eye movements and some leg movements through-out all stages and sometimes I go to my computer to work while still officially being asleep. But I don't remember any of it. Kristian, how-ever, can usually figure out when I've been sleep-working, because the writing style slightly changes.

The REM sleep cycle holds the key to being able to solve prob-lems or increase creativity while we're asleep because our brain waves during this stage lie in the same range as the brain waves when we are awake.

Controlling Your Dreams

When I went to graduate school, lucid dreaming was a concept every-one knew of, yet knew nearly nothing about. Generation X missed the lucid dreaming debates of the 1960s, 1970s and 1980s. After that the debates about lucid dreaming faded out and it became the geeky sub-ject matter of a few liberal intellectuals hardly anyone had heard of. Christopher Nolan's movie *Inception,* about mind thieves who hack into people's subconscious minds to steal their secret ideas, brought the concept of lucid dreaming back into the minds of the masses.

Though it appears to involve lucid dreaming, *Inception* is not the best example of the phenomenon as it actually occurs in real life. In most lucid dreams, you don't have car chases and gunfights. Also in most lucid dreams, you don't enter dreams folded inside other dreams. But the movie successfully illustrates some core elements of lucid dreaming, for example, the possibility of manipulating dream content.

Lucid dreaming is your chance to play around with the extraordi-nary abilities buried in unused parts of your brain. Regardless of whether you are superhuman in waking life or not, you can put the deepest areas of your brain to good use while you're sleeping. All the obstacles of reality can be set aside, as you make trips to the sun or the interior of the earth, find a way to test neural transmission using a new experimental design, or hit upon an idea for a new generation of the Segway scooter.

A lucid dream is one in which the dreamer realizes that he or she is dreaming. Many times, the dreamer is able to control his or her dream, but that isn't always the case. Compared to regular dreamers, lucid dreamers show more activity in their conscious control center in the prefrontal cortex while the rest of the brain is in a dreamlike state. In other words, in a lucid dream, your brain's control center is awake while the rest of the brain is asleep.

In terms of consciousness, lucid dreaming is a hybrid state that that is characterized by both waking and dream consciousness. Lucid dreaming is biochemically similar to ordinary dreaming. They are characterized by similar levels of neurotransmitter release. They are also phenomenally similar, that is, they share the same what-it's-like-ness. They both involve visual imagery. But consciousness accompanying lucid dreams lies in the higher activity level of the frontal areas of the brain. This is also a feature of very vivid and active states of waking consciousness, such as deep concentration and active listening. But the fact that you're not living in reality means that lucid dreaming differs from consciousness experience in one dramatic way: You can learn to control the narrative of your own dreams.

Psychologist Robert Waggoner, author of *Lucid Dreaming: Gateway to the Inner Self*, is the former president of the International Association for the Study of Dreams (IASD). He was one of the early promoters of lucid dreaming, recommending it not just for entertainment but also for the treatment of depression, anxiety, phobias, nightmares, and post-traumatic stress disorder. According to Waggoner, being able to control your dreams depends, to a high degree, on your expectations. You can bounce into a wall and feel pain but only if you expect to feel pain. If you decide and genuinely expect not to feel pain, then you don't feel anything. So you can sort of turn on and off your senses.

So how do we figure out that we're dreaming? Certain features of dreams can tip us off. They vary from person to person, but there appears to be some overlap. Clocks and watches tend not to work reliably in dreams. You normally cannot read a letter twice, as it will say something different on the two occasions. Change blindness is prevalent in dreams: The colors of people's shirts quickly shift—sometimes people show up at a party in one outfit and later they are wearing something entirely different. If you ask them, they will always have a story to tell you. But the story doesn't make sense. For example, they will say that they brought the second outfit but you noticed they didn't carry anything with them when they entered. People who were not in the room at the beginning of the event suddenly materialize, small rooms suddenly are large ballrooms, the establishment suddenly has a kitchen, food magically appears. You can train yourself to pay attention to these kinds of changes. Unexplained changes are signs that you are dreaming. The way to acquire knowledge that you are dreaming while you are dreaming is straightforward: You gather evidence from the dream in support of your belief that you are dreaming. When you have collected sufficient evidence, you know that you are dreaming.

How to Dream Lucidly

The first step in maximizing your dreaming potential is to make sure you're getting an adequate amount of sleep. But what is a good night's rest? To some extent it depends on the individual. Some successful people like comedian Jay Leno, Yahoo CEO Marissa Mayer, and fashion icon Tom Ford claim to need as little as four hours of sleep per night. But that isn't normal. In fact, after a few four-hour nights, most people can expect to start seeing the signs of sleep deprivation. It ain't pretty. You'll start out by being excessively tired, interrupting your own sentences with sustained periods of yawning. You'll fight to keep your eyes from shutting on their own. After a while, you'll start hallucinating, forget entire routines. Your brain will be a mess. It should come as no surprise that sleep deprivation can be a very effective method of "extracting information" from "enemy combatants."

The average adult needs around seven to eight hours of sleep a night. You need a bit more at birth, and less and less as you continue to age. The exact amount one needs depends on that individual's circadian rhythm, a bodily cycle that repeats about every twenty-four hours and regulates everything from hormone release to body temperature and sleep cycles. The rhythm is entrainable, meaning that it changes in response to environmental stimuli. Humans deprived of light exposure will see their circadian rhythms start to drift in length. It's also highly susceptible to the influence of chemicals such as caffeine and alcohol. Despite how it might seem, you can't get a good night's sleep when intoxicated. Although alcohol consumption typically leads to fatigue, it also disrupts your circadian rhythm. Not only does it adversely affect REM sleep, it causes your sleeping brain to "rebound," meaning that it actually tends to prevent you from reaching a meaningful level of sleep.

Your ability to sleep can also be affected by your diet. Using a national survey of over four thousand people, researchers recently looked at a myriad of diet traits to determine if good sleepers have something in common with one another. They first divided the participants into three groups depending on the number of hours of sleep they reported: Very short sleepers got less than five hours per night, short sleepers got five to six hours of sleep, standard sleepers got seven to eight, and long sleepers got nine or more hours of sleep a night. Not surprisingly, those who regularly consume adequate amounts of water along with a balanced diet tend to get their full eight hours. Very short sleepers consume the most calories, drink less tap water, and consume less carbohydrates and lycopene, a chemical found in red and orange fruits. Short sleepers consume less vitamin C, water, and selenium, a chemical from nuts, meat, and shellfish, than standard sleepers, but consumed more lutein and zeaxanthin, found in leafy green vegetables. Long sleepers eat less theobromine, found in tea and chocolate, a saturated fat called dodecanoic acid, and choline, found in eggs and meat. So if you're having trouble staying asleep, it might be prudent to adjust your diet accordingly.

Good sleep is also highly dependent on your environment. This might come as a surprise if you're one of those individuals who can sleep through almost anything. Temperature is one important aspect. Studies indicate that an ambient temperature is best around 61–66 degrees Fahrenheit, so long as you use at least one bedsheet. Your environment should be quiet as well. Noise above fifty decibels reduces sleeping time. That's about the same volume as a quiet conversation. A cell phone alert chime is certainly louder, so leave those things on vibrate. And the amount of ambient light has also been shown to greatly affect the timing of sleep cycles. Futurists predict that sleep's fickleness will spur the next series of hotel wars. Some hotels are already researching customized sleep experiences that can adjust everything from ambient temperature to the amount of light entering the room, the firmness of the mattress, and even the cushiness of the pillows. All of this will bring you the luxury of something you won't be conscious to enjoy. How ironic.

Once you get your sleep patterns in order, it's time to train yourself to get better at recalling your dreams. All the provisions required to reach lucidity won't be worth it if you can't remember enough about your dreams to know they're working. There are two tricks experts agree will make you better at remembering. First, before you go to sleep, consciously tell yourself that you intend to remember your dreams. As you fall asleep in the dark, tell yourself over and over "I will remember my dreams." Second, get in the habit of reflecting on the dreams you spontaneously do remember. When the memory of a dream pops into your head, really study the imagery in your mind. Think about how the choices you made in your dream differ from what you'd normally do. Reflect on the wacky details that violate the laws of physics. Imagine how the story line could have developed differently. If your mind is particularly stubborn about remembering, it helps to keep a dream journal. Whenever you wake up from a dream, write down as much as you remember. Then study it in the morning. You'd be surprised how much it will help you get better at remembering. Once you start remembering your dreams, your mind is ready to start dreaming lucidly.

Lucid dreaming can occur automatically, but usually it's an art. It's something that requires training. Robert Waggoner started lucid dreaming when he was a junior in high school in 1975. Up until then he had been really interested in dreaming. He had always paid attention to his dreams. In 1975 Waggoner read a book by Carlos Castaneda called *Journey to Ixtlan*. It suggested you look at your hand before you go to sleep and tell yourself "It's not my dream. It's not my dream." You just keep repeating this until you fall asleep. The key is that your hand becomes so ingrained in your mind that it starts popping up in your dream at inappropriate places. That triggers your realization that you're dreaming. In high school Waggoner did it so consistently that his hand would show up in his dream almost every night. He would be walking through his high school hallway and then boom, his hand would be right in front of his face. He thought, "Oh God! It's my hands. This is a dream." Since then he has practiced lucid dreaming and is now what we could rightly call the world's leading expert in the subject.

In 1985, psychophysiologist Stephen LaBerge invented a technique called the Mnemonic Induction of Lucid Dreams (MILD). It's implemented at times when you wake up from a dream in the middle of the night. You recall whatever you can of the dream you just had. Then you tweak it, so that in your remembered version you become lucid at some appropriate point. Then you make yourself go back to sleep and tell yourself, "When I fall asleep I will go back to realizing the dream. When I start dreaming I will realize that it is a dream." Oftentimes you'll find yourself picking up where you left off, but with an added bonus of lucidity.

There are other ways to develop a lucid mind. For example, you can form the habit of examining the environment or your state of awareness during the day. Mental habits you practice during the day tend to continue in dreams. So you examine your environment during the day, you examine your awareness, and then you may notice that something is different once you start dreaming. For example, you might stop in the middle of your dream and think, "Hey, it's three a.m. and the television is on. I distinctly recall turning it off before I went to bed. Something is wrong here. Let me grab a book from the bookshelf. Aha! I can't read the book. This must be a dream."

You are not actually in a lucid dream just because you look at your watch in your dream in order to check whether you are dreaming lucidly. But when you succeed in noticing that, say, your watch isn't showing the time correctly or that you can't read the same sentence twice, *and* you conclude that you are dreaming, you have entered a lucid dream state.

Perhaps the best way of achieving lucid dreaming is to make a habit out of checking clocks twice. The second time you check you may not be able to see what time it is, or the clock may show something ridiculous, such as 69:38. That should make you realize that you are dreaming. "So I hope that someone won't trick me one day with a

broken clock or something like that," Waggoner jokes. But it's a common way to achieve lucid dreaming. People notice things that don't
work, for example, clocks, the television or even your own eyelids.
Those are clues that something is going on that doesn't go on outside
of dreams.

But the tipoff can also be something very simple. Waggoner reports that he sometimes sees light reflecting in the middle of nowhere,
which makes him think "Wait a second, that doesn't look right. This
must be a dream. Or I see a late family member and I say, 'wait a
second that person is dead. Oh, yes, this must be a dream.'" Waggoner
also recommends taking the dietary supplement glutamine to make it
easier to enter a lucid dreaming state. Glutamine is one of the twenty
essential amino acids found in protein-rich foods. In 300–1000 mg a
day, it may help you stay focused, he says.

Once you have become aware that you are dreaming, the trick
now is to start taking control over your dream actions. Suppose you
have succeeded in taking control on a couple of occasions. The things
you want to do are not always possible. Say I want to have a conversation with the philosopher Saul Kripke. I won't necessarily be in a position to find Kripke in the dream. So, having a conversation with him
may be impossible even inside a dream. I might succeed in finding
him but then might experience difficulties getting him to talk. There
are some things you can do to help this process along. If you want
certain people to be present in your dream, it helps to think about
them a lot before you go to sleep. It also helps imagining the activity
you would like to engage in with them before falling asleep. That will
make it more likely that your brain will create dream stuff of a desir-
able kind for you to interact with.

Waggoner tells us that lucid dreams follow their own dream laws.
"You can't stare at something fixed too long," he says. "If you do, it
starts shaking and you wake up." If you are about to wake up, the best

thing you can do is to look around. Stephen LaBerge thought that spinning may stimulate the dopamine system that keeps dreaming going. LaBerge is famous for having invented several devices to help people enter lucid dream states, among others, the lucid dream mask. The lucid dream mask uses infrared technology to detect the rapid eye movements that occur during REM sleep. It then flashes lights that pass through your eyelids and makes sounds that you can pick up and integrate into your dreams. The lights and sounds from the outside world are supposed to help you stay in the lucid state and be able to control the dream. It is like a voice with power that comes to you in your sleep saying, "You're only dreaming. Now take control."

Lucid Dreaming, Knowledge, and Creativity

Lucid dreaming can take you to incredible places: Machu Picchu, the interior of the earth, or a Grateful Dead concert in Oakland. But it's not all fantasy. Sometimes lucid dreams allow you to gain access to implicit knowledge that exists in your brain, but that you don't have access to in waking life. You can interact with your unconscious self in a lucid dream. Lucid dreams can reveal a deeper layer of the self, another layer that seems to contain more information about the real conscious self than the waking ego.

Waggoner tells us that he knew an author who told him that whenever he got stuck writing his novels, he could solve it in his lucid dreams. He would consciously call the novel characters to him and ask what was wrong. And one of the characters of the novel might say, "Oh, you shouldn't kill Mary in chapter 3, because we need her in chapter 12 to resolve everything." He also met a professional musician who said that he became lucid and could hear rock lyrics, which he would write down the next day. "So people are using this in a lot of different ways in some kind of interaction with their inner muse and their inner creativity," Waggoner says. One artist says that all of his paintings come from lucid dreams. Before he begins a painting, he dreams that he's standing at a door that separates him from his next masterpiece. He opens the door and finds a piece of art on a wall, which he then studies for a long time. When he wakes up, he paints it from memory.

It is well known that dreams sometimes are the place where problem solving takes place. But that doesn't necessarily mean you must be lucid for dreams to result in new discovery. Many famous and influential ideas originate in dreams. The Danish physicist Niels Bohr hit upon his famous model of the atom during a dream, Albert Einstein thought of the speed of light as a constant after reflecting on a dream in which he was hurling down a mountainside at the speed of light, the German chemist Friedrich August Kekulé discovered the benzene ring in a dream about the molecules taking the form of a snake biting its own tail, the Spanish painter Salvador Dali's melted clock painting called *The Persistence of Memory* came from a dream, and Vincent van Gogh used to say "I dream my painting, and I paint my dream."

Lucid dreaming isn't just for solving puzzles. It can also be used as therapy. For example, kids with frequent nightmares can be taught to get out of their nightmares through lucid dreaming. When taught to become lucid, they can "step back" and think to themselves "Right. It's just a dream."

This worked for a woman who had lost a leg. She had constant nightmares and was told to use lucid dreaming to get rid of them. She had a dream that she was running away from a monster. But after becoming lucid, she thought to herself "Wait a minute! I have one leg, how could I be running?" She turned to face the monster and destroyed it. The nightmares ceased after that. And she was subsequently thrilled to be able to experience running again in her lucid dreams.

In fact, lucid dreaming can be very effective in getting rid of fears and phobias. It works just like cognitive behavioral therapy (CBT), a type of treatment employed by many psychologists for severe obsessive-compulsive behavior, among other conditions. The idea behind CBT is that the best way to get over fear of a particular situation is to place yourself in that situation over and over again until your brain becomes desensitized to the trigger. So, someone with a fear of heights would imagine being in a high spot, such as the top of the Seattle Space Needle, and imagine themselves feeling calm there. After a while, the patient might progress to *actually* putting themselves in safe but scary situations. Over time, patients typically become desensitized to the trigger. Lucid dreaming can be used in a similar way. Waggoner says he once cured a woman's fear of flying. He taught her lucid dreaming, then told her to start getting on planes in her dreams. At the beginning she would be equally fearful when riding on the dream planes. But over time she overcame that fear without being on a single real plane. Now she is flying all the time in real life.

People can even use lucid dreaming to get over bad breakups or bereavement. In lucid dreaming therapy for getting over the loss of loved ones, the trick is to find the loved one in the dream and have those final conversations with them that keep plaguing the mourner in waking life. But it can also be psychologically dangerous to meet lost loved ones in dreams. It's tempting to continue to have a dream relationship with the person, which interferes with the attempt to get over the loss.

Using lucid dreaming as therapy, of course, is in no way straightforward. Changing dream content at will is exceedingly difficult. While you have some control over your own actions, it is difficult to control what other people do. It is difficult to make their behavior deviate from their normal waking behavior. If you ask your tenth-grade schoolteacher to undress in the middle of the street, she is not going to do it. She might smack you. Failures to manipulate dream content often are grounded in a desire to make big changes all at once. Your brain is not going to allow that. You have to begin with little changes. If you have a desire to see your teacher in her undies, you should probably start hanging out with her in your dreams first. The brain needs a coherent narrative. Even though it will force abrupt changes on you, it will not let you force abrupt changes on it.

As you progress through your journey, you'll find it easier to maintain lucidity, allowing you to perform much more complicated tasks with great efficiency. Our minds might indeed be functioning more efficiently during dreams than in real life. In dreams our minds appear to be more like Jason Padgett's and Derek Amato's, perhaps allowing more of the complicated calculations and attention to details to take place below the level of conscious awareness.

CHAPTER 8

Super-Perceivers

The Multifaceted Ways of Experiencing the World

Daniel Kish has been blind since he was thirteen months old, but you wouldn't be able to tell. He navigates through crowded streets on his bike, camps out far in the wilderness, swims, dances, and does many other activities many would think impossible for a blind person. How does he do it? Daniel is a human echolocator, a real life Daredevil. Using a technique similar to bats' and dolphins', human echolocators navigate based on audio cues given off by reflective surfaces in the environment. Few people are aware that this same technique can work for human beings. But as a matter of fact, echolocation comes quite naturally to people who are deprived of visual information like Daniel. "I don't remember learning this," Daniel says. "My earliest memories were of detecting things and noting what they might have reminded me of and then going to investigate."

Daniel was born with bilateral retinoblastomas, tiny cancers of the retina, which are part of the eye responsible for sensing visual data. In this type of cancer, tumors begin to form early, so aggressive treatment is necessary to ensure it doesn't metastasize to the rest of the body.

Unfortunately, the tumors cannot be separated from the retina. Laser treatments are performed to kill them off, followed by chemotherapy. The result is that the retina is destroyed along with the cancer. This means that patients often are left completely blind. Daniel lost his first eye at seven months and the other at thirteen months. He has no memory of having eyesight. His earliest vivid memory is of standing by a chain-link fence in his backyard. He was very young, maybe two-and-a-half. At about midnight, he climbed out the window of his bedroom. He stood over by the fence, angled his head upward, and clicked over it with his tongue, listening for the echo. He could tell that there were things on the other side. Curious as to what they were, he climbed over the fence and went over to investigate. That night, he climbed multiple fences into multiple yards looking for what was around.

Like Daniel, Ben Underwood was a self-taught echolocator. Ben Underwood was also diagnosed with bilateral retinoblastomas, in his case at the age of two. Following many failed attempts to save his vision by treating the tumors with radiation and chemotherapy, his mother made the difficult decision to remove the boy's right eye and remove the left retina. This left Ben completely blind. His mother never wanted him to believe he couldn't fare as well as everyone else in the world. "Mommy, I can't see anything," Ben cried when he woke up from the surgery.

His mother had tears in her eyes as she instinctively replied, "That's not true, Ben. You cannot see with your eyes. But you can hear me. And you can feel me." Mrs. Underwood took Ben's little hand and put it on her face. "So you may not be able to see with your eyes but you can see with your ears and you can see with your hands. You can see, Ben." A couple of years later when Ben was on the backseat of the car with the window down, he suddenly said, "Mom, do you see that tall building there?" Shocked by the boy's statement, the mother responded, "I see the building, but do *you* see it?"

It turned out that Ben had picked up on the differences in sounds coming from empty space versus a tall building. When Ben was in school, he started clicking with his tongue, first as an idle habit, but then he realized that he could use the skill to detect the approximate shape, location, and size of objects. Soon Ben was riding a bicycle, skateboarding, playing video games, walking to school, and doing virtually anything else an ordinary boy his age could do. Ben never used a guide dog, a white cane, or his hands to navigate. Very sadly, Ben passed away in 2009 after the cancer that claimed his eyes returned.

The Phenomenology of Echolocation

For centuries, researchers have been trying to find out how blind people compensate for their loss of vision. Everyone knew that some blind people occasionally were able to "hear" objects that were apparently making no sounds. But no one knew exactly how they were able to do this. Echolocation in bats was documented in 1938. But scientists didn't become seriously interested in the phenomenon until the early years of the Cold War, when military funding made the research feasible. Until recently, echolocation had not be been thought to be a kind of vision. Human echolocation is akin to active sonar and the kind of echolocation employed by dolphins and bats, but less fine-grained. While bats are able to locate objects as small as flies, human echolocators say the objects must be much larger—about the size of a water glass.

Philosophers and neuroscientists often talk about *phenomenology*, that is, "what it's like" to have an experience. If we show you a red ball and ask you about its color, assuming you're not color blind, it should be easy for you to answer "red." Computers can do that too. However, if we ask you to describe *what it's like* to see the color red, you'd have a much harder time giving us an adequate answer. By their very nature, questions about phenomenology can be nearly impossible to answer. This makes it hard to discover exactly what it's like to experience echolocation. American philosopher Thomas Nagel became famous for arguing that humans shouldn't ever be able to understand the phenomenology of echolocation. At least from our viewpoint, there's nothing it's like to be a bat. Though many philosophers raised principled objections to this claim, it turns out that the best objection comes in the form of an actual counterexample.

It's hard to determine whether echolocators experience visual imagery in response to sound signals. Daniel says that because he has been blind for as long as he can remember, he has nothing to compare his experience to. He cannot really say whether his experience is like seeing. However, he says he definitely has spatial imagery, which has the properties of texture and dimension. Research indicates that the imagery of echolocation is constructed by the same neurology that processes visual data in sighted people. The information is not going down the optic pathway—the connection from the eyes to the brain—but it ends up in the same place. And some individuals who have gone blind later in life describe the experience as visual, in terms of flashes, shadows, or bright experiences. It seems very possible that echolocators have visual imagery that is similar to that of sighted people.

Ben's case provides some evidence that echolocators form visual images of the shape and size of objects in their surroundings. Ben consistently reported that he was seeing the objects he could detect, not just hearing them. Self-reports, of course, are notoriously unreliable. However, there is other evidence that Ben really could see with his ears.

Ben had prosthetic eyes that took the place of his real ones. Because his eye muscles were still intact, the prosthesis would move in different directions much like real eyes. In the documentary, *Extraordinary People: The boy who sees without eyes*, Ben's prosthetic eyes can be observed making saccades in several situations that require focusing quickly on different objects in the peripheral field. Saccades are the quick, coordinated movements of both eyes to a focal point. The documentary never discusses Ben's saccadic eye movements but there is no doubt that they occur, matching the auditory stimuli he receives. For example, in one scene he is playing a video game that requires destroying objects entering the scene. Though echolocation couldn't give him the ability to "see" the images on a flat screen, Ben was able to play games based on the sound effects played over the television's speakers. Like many blind people, Ben was able to use the sound cues to figure out where objects were on the screen. The saccadic eye movements corresponded to the changes in location of the virtual objects.

The primary role of saccadic eye movements is to guarantee high resolution in vision. We are capable of seeing with high resolution only when an image from the visual field falls on the central region of the retina, called the fovea. When it falls on the more peripheral areas of the retina, we don't see them very clearly. Only a small fraction of an entire scene falls on this region at any given time. But rapid eye movements can ensure that you look at many parts of an entire scene in high resolution. Although it is not immediately apparent, the brain creates a persisting picture on the basis of the many individual snapshots.

Many other factors govern eye movements: changes in the environment, beliefs about the environment, and intended action. For example, a sudden noise can make your eyes move in the direction of the noise. A belief that someone is hiding in the tree in front of you can make your eyes seek out the tree. And intending to climb a tree can trigger your eyes to switch from the ground to the tree so you can inspect it properly.

When recalling visual imagery, there is no sensory input or external environment to be inspected. However, saccadic eye movements still occur even when you are just remembering what something looks like and not actually looking at it. This happens because when the brain stores information about the environment, it stores information about eye movements together with that information. When a visual image is generated, this is likely a composite of different snapshots of reality. Rapid eye movements help keep the visual image organized and in focus. This explains Ben Underwood's saccadic eye movements: The sound stimuli from Ben's environment triggered his brain to generate spatial imagery matching the sound stimuli and the saccadic eye movements helped keep the image organized and in focus.

Seeing with Your Ears

Sighted people often use a simple form of echolocation too, perhaps without even realizing it. When hanging a picture, one way to determine where a stud is located within that wall is to knock around and listen for the changes in pitch. Carpenters do this all the time. Blacksmiths bang on metal in search of imperfections. When you tap on a hollow space in the wall, you usually don't hear an actual echo—yet you can tell somehow that the space sounds hollow. Research shows that we can perceive these types of stimuli unconsciously. When we do hear echoes, it's from sound bouncing back off distant objects. When you click your tongue or whistle toward a nearby object, the echo returns so fast that it overlaps the original sounds, making it hard to hear an echo. But the brain can unconsciously detect the combination of the sound thrown in one direction and the returning sound as an alteration in pitch. What makes Ben and Daniel so remarkable is that they can use what everyone's brain unconsciously detects as a way to navigate the world. So we have here again a case in which being able to act on unconscious brain processing is accountable for supermind abilities. Research confirms that echolocation is also a potential ability of sighted human beings. After all, the visual cortex processes some sounds in all of us, particularly when the brain seeks to match up auditory and visual sensory inputs.

American psychologist Winthrop Niles Kellogg, who is best known for a controversial experiment in which he raised a baby chimpanzee alongside his own infant son, began his human-echolocation research program around the time of the Cuban Missile Crisis. His research showed that both blind subjects and sighted subjects wearing blindfolds can learn to detect objects in the environment through sound. A study by another researcher showed that with some training, both blind and sighted individuals are capable of precisely determining the properties of objects such as distance, size, shape, substance, and relative motion. While sighted individuals show some ability to echolocate, blind echolocators seem to operate a bit differently when collecting sense data. Kellogg showed that blind echolocators move their heads in different directions when spatially mapping an environment. Sighted subjects don't move their heads when given the same types of tasks. Another study demonstrated that sighted individuals are better able to echolocate if they are allowed to move.

We know that blind individuals often experience enhanced perception in nonvisual sensory modalities like sound or touch. One study found that touch and hearing can give rise to increased brain activation in the visual cortex in blind people. The visual cortex was not activated by the mere presence of sounds but became involved when the brain needed to shift attention to a change in the incoming sound stimuli. This suggests the visual cortex assists the sense of hearing with action orientation. The ability to detect these sorts of changes may have had the evolutionary benefit of helping the individual avoid potentially dangerous situations.

Sighted subjects who are deprived of visual sensory information for an extended period of time naturally start echolocating, possibly after only a few hours of being blindfolded. And their echolocation is associated with visual imagery. After a week of being blindfolded the hallucinations become more vivid. One research participant said that he experienced "ornate buildings of whitish green marble and cartoon-like figures." The subjects in Alvaro Pascual-Leone's study were also examined while performing tactile and spatial discrimination tasks in a functional MRI. The brain scans showed that the neural activation in the visual cortex when using touch to discriminate among different objects increased in subjects over the course of the study. While the visual cortex seemed to be recruited more, scans also indicated the somatosensory and auditory cortices were recruited less.

One might wonder if the changes in brain activity in some blind individuals are due to restructuring or rerouting, where restructuring involves generating new neurons or new connections between existing neurons. Research points to the latter. There are two details that suggest that the brain automatically responds to sensory deprivation by utilizing data from other senses. First, the short duration of the Pascual-Leone study coupled with quick changes in people's behavior indicate that the brain did not have time to make new neural connections. So sound information is likely rerouted when people are blindfolded. Second, the hallucinations and visual cortex hyperactivation ceased when the blindfolds were removed. This suggests that echolocation may be accomplished by simply activating the right neural structures after sensory deprivation. From these findings, Pascual-Leone et al. argue that the visual cortex is specialized for spatial discrimination. While it prefers visual input, it can utilize information from the other senses to generate spatial information.

Is sound alone responsible for echolocators' ability to navigate the environment? Researchers wonder whether other touch cues are responsible. For example, it is possible that the way air moves around objects can offer cues about the surroundings. Philip Worchel and

Karl Dallenbach from the University of Texas at Austin sought to answer these questions in the 1940s. Their experiments involved asking
both blind and blindfolded sighted participants to walk toward a
board placed at varying distances. Participants were rewarded for
learning to detect the board by not walking into it face-first. After
multiple trials, both the blindfolded and blind participants became
better able to detect the obstacle. After about thirty trials, blindfolded
participants were as successful at stopping in front of the obstacle as
blind participants when they were wearing hard-soled shoes. This
ability disappeared when participants performed the same experiments on carpet or while wearing socks, decreasing the amount of
sound created in the room. The researchers concluded that the participants relied on sound emanating from their shoes. So, apparently
sound is responsible for navigational ability.

The increased ability to navigate based on sound appears to be
the result of sound processing in the brain, not merely increased acu-
ity of hearing. This suggests that echolocation is being accomplished
neurologically. One study showed that the blind and the sighted scored
similarly on normal hearing tests. But when a recording had echoes,
parts of the brain associated with visual perception in sighted people
became active in echolocators but not in sighted people. This showed
how echolocators extracted information from sounds that was not
available to the sighted controls.

Conscious reports of subjects have been thought to show that humans aren't able to perceive objects that are very close—within two
meters of them—through echolocation. But a 1962 study by Kellogg
at Florida State University showed that blind people are able to detect
obstacles at much shorter distances—30–120 centimeters. Impres-
sively, some of the participants were accurate within 10 centimeters.
This suggests that while subjects aren't consciously aware of the echo,
they are still able to respond appropriately to echo stimuli. At these

distances, the delay between the generation of a sound and the returning echo is only 0.3 milliseconds. Even though this acute sensitivity is amazing, it pales in comparison to that of bats, who can detect delays of approximately 10–12 nanoseconds. Oddly, that is faster than the action potential of neurons, leaving the mechanism in need of explanation.

Precision timekeeping abilities may be enhanced in those who naturally echolocate. Another echolocator, Ellen, as a child used to run through the forest behind her home, never hitting a single tree in spite of her blindness. But in addition to being an echolocator, she is also a prodigious savant, a human music recorder. Ellen is able to recall and then replay with extreme precision any song she has ever heard. At six months of age, her sister observed her humming the tune of Brahms's "Lullaby," the song that emanated from her cradle gym as she fell asleep. Ellen started playing music at age four, but performing became her life at age seven when her teacher convinced her parents to buy a piano. Not only can she replay music heard only once with startling precision, she can also improvise in real time, playing the same song in the style of jazz, ragtime, and then classical.

Ellen also appears to be able to keep perfect time. Before she was ever formally taught to tell time, her mother used to play the automatic time recording on the telephone to help her get over what appeared to be an intense fear of the telephone. After listening for ten minutes one day, Ellen fully understood how the numbers in time sequences advanced to represent the different periods of the day. This is likely related to her musical savantism, since music is inherently time dependent. Because the ability to track time with precision is needed for the brain to compare the echo signals coming from objects in the environment, Ellen's brain likely organized itself in such a way to enhance her timekeeping ability. It is possible that her musical savantism is a wonderful side effect of that enhancement.

The above cases show that our perceptual experiences involve a lot more than just what we are consciously aware of. Our brain is primed to accomplish the seemingly superhuman even at the level of basic perception. It has the exceptional task of turning waves merely striking the eardrum into complicated phenomenal representations of the surroundings. It responds to sensory deficits in a way that seems almost intelligent, restructuring itself to make up for its limitations. The brain is good at creating our rich experience, so good in fact that many wonder why we have experience in the first place.

Teaching the Blind to See

When seeing Daniel interact with the environment, it's easy to forget he has no use of his eyes. In fact, we initially found ourselves a little irritated when first interviewing Daniel over Skype. We had spent a considerable amount of time rearranging the background of our office to make it look like we're very organized and neat. But upon connecting with Daniel, we noticed the video feed was turned off. After about a half-hour of internal complaining about Daniel's lack of consideration, we had an epiphany—Daniel is *blind*. Duh! This goes to show how Daniel's confidence makes it easy for us to forget about his affliction.

Perhaps what is most amazing about echolocation is that it can be taught. Daniel wrote his developmental psychology master's thesis on the subject, creating the first systematic approach to teaching echolocation. Now through his organization, World Access for the Blind, he strives to give the blind nearly the same freedom as sighted people by teaching blind people to navigate with their ears.

So why don't all blind people echolocate? The problem, says Daniel, is that our society has fostered restricting training regimens. One example of this restriction is the traditional method of cane training. The techniques behind using a cane were developed by the military over sixty years ago for blind veterans, people used to living under many restrictions. For them, it was quite easy to adapt to the regimented system of using the cane. But now this system is taught to everyone, including children and the elderly.

Daniel's training curriculum differs from tradition by taking an immersive approach intended to activate awareness of the environment. It's a tough-love approach with very little hand-holding. Children are encouraged to explore their home environment for themselves while family members are discouraged from interfering unless the child otherwise could be harmed.

Daniel then changes the cane used during walking. The traditional cane's length is supposed to be sternum height. The person holds the cane out in front, elbow slightly bent, so the hand is roughly at waist height. With every step taken, the cane tip lands about where the person's foot will land so that the cane clears the area of ground where she is about to step. The person repeatedly taps the cane from left to right, called "two-point touch." Though this method is supported by computer models, Daniel believes the movement is quite unnatural: "We are not robots. The reality is that the biomechanics do not sustain the kind of regimented movement you have to have in order for that to work—you lose fluidity of motion. You don't have to be a physical therapist to know that's a recipe for a wrist problem." Daniel trains using a cane modified to be slightly longer in length.

Then the real fun begins—the students learn to echolocate by systematic stimulus differentiation. Notice the term "detection" isn't used. No stimulus really occurs in a vacuum, so the process is not so much detection as it is distinguishing one stimulus from another and its background. The process follows a standard learning structure—students first learn to differentiate among strong, obvious stimuli and then advance to weaker, less obvious stimuli. Daniel establishes a "hook" stimulus by using a plain panel he moves around in the student's environment. Students don't need to know what they're listening for, since the stimulus is selected to be strong enough that it captures the brain's attention. But the panel doesn't work for everyone at first. In those cases, he uses a five-gallon bucket or something else that produces a very distinct quality of sound. Once the brain is hooked on the characteristic stimulus, he starts manipulating its features to make the effect subtler.

The next set of exercises helps students learn how to determine what the objects actually are. This essentially involves three characteristics—where things are, how large they are, and depth of structure. A scientific term, "depth of structure" refers to the geometric nature of the object or surface. Students are asked to answer questions such as "Is the object coarse or smooth?", "Is it highly solid or sparse?", and "Is it highly reflective or absorbent?" Kish says all of those patterns come back as acoustic imprints. The key is to notice the changes in the sound when it comes back from when it went out. With determined practice, students eventually learn how to differentiate among environmental stimuli in general.

Though the organization's instructors are currently all blind echolocators, Daniel anticipates that sighted instructors can teach the skill as well. Several sighted instructors are currently in training, showing promise of being able to use echolocation themselves. His goal is to have sighted instructors performing just as well as blind echolocators, though he emphasizes that although both sighted and blind individuals can echolocate, they may have profoundly different phenomenological experiences. Brain imaging may give us a clue. In the meantime, Daniel is hard at work, teaching the blind how to use their ears to see.

Music to His Eyes

Now imagine hearing what you see. This is reality for Lidell Simpson, a vision-to-sound synesthete. "Take this example of observing a single light being turned on then off," Lidell explains. "I will hear the 'plink' of the turning on, I will hear the 'whine' of the light and finally the 'plunk' of the light turning off." This description of his synesthesia is amazing considering that Lidell was born profoundly deaf in both ears.

For more than five years, doctors mistakenly told Lidell's parents that any hearing assistance could permanently damage the little hearing that Lidell might manifest at a later age. Because of the difficulty in communicating with Lidell, doctors assumed he was mentally handicapped. Thanks to a caring family and despite the doctors' repeated insistence that Lidell be institutionalized, he eventually was able to start understanding people through lip reading and with the aid of an electronic device that greatly amplified sounds. After years of speech therapy, he learned to speak. It quickly became clear that Lidell was an intellectually gifted individual. In college he became interested in computer programming and later worked for many top IT firms. Over the years Lidell has become acquainted with respected researchers in neuroscience, philosophy, and psychology who regularly call on his experiences to contribute to research.

Lidell is not only intellectually gifted. Like Jason and Derek, he also has what may appear to be superhuman abilities. Lidell can detect features of sounds he cannot consciously hear when required to make a guess. This is also known as "deaf hearing." What's perhaps most interesting about Lidell is that he hears a unique "ping"—a sound—corresponding to every face he recognizes. If Lidell quickly scans a crowded room, he will hear a "ping" corresponding to the locations of people he has met before. He once was distracted by a "ping" coming from oncoming traffic on the freeway—and saw a family friend driving by in the opposite direction.

Lidell has what's known as mixed hearing loss, meaning that there are multiple conditions causing his deafness. Sensorineural hearing loss comes from an impairment of the inner ear, an impairment of the nerve that transmits impulses from the cochlea to the hearing center in the brain, or an impairment of the part of the brain involved in interpreting sounds. Conductive hearing loss is an impairment of the inner ear, the part of the ear that precedes the cochlea. This kind of deafness prevents the sound from reaching the sensory apparatus. Lidell has profound sensorineural hearing loss (the worst kind short of complete deafness), and severe conductive hearing loss (the second to the worst kind).

Together with bestselling author and neuroscientist David Eagleman at Baylor College of Medicine in Houston and Janina Neufeld at the University of Reading, our synesthesia lab conducted several studies of Lidell. In a brain imaging study of Lidell we found that the part of his brain that is involved in hearing (the superior temporal gyrus on the side of the head, where the golf ball-sized sound area known as the auditory cortex is located) was smaller than average for both hearing and deaf people.

Even people who are profoundly deaf can hear some sounds if they are loud enough or if they have a particular frequency. Without hearing aids, Lidell can hear certain high-frequency sounds such as the clapping of a hand or a hammer striking a nail at close range. But the sounds that reach Lidell's brain are meaningless to him. He cannot hear the clapping of a hand *as* clapping or a hammer striking a nail *as* a hammer. For Lidell these sounds do not have a source. They lack location and direction; they belong to no one. They are like quiet background noise, like a fan in the corner of a noisy room.

The brain processes that underlie hearing are similar to those underlying sight. The hairs in the inner ear can be compared to the retina in the eye. The hairs are attuned to different frequencies, whereas different parts of the retina are attuned to differences in location. The information from the retina is further processed in the deeper parts of the brain and then passed on to the primary visual cortex in the back of the head, a brain region that processes what you see. Sound information from the inner ear is likewise processed in the deeper parts of the brain and then passed on to the primary auditory cortex in the temporal lobe—the area that processes what you hear.

Studies of people with damages to the auditory cortex show that auditory information must reach the auditory cortex in order for people to be conscious of sounds. People with severe lesions to the auditory cortex are deaf: They cannot consciously perceive sounds. This is analogous to the case of vision. When people have damages to the primary visual cortex, they cannot consciously see anything. Everything is pitch-dark or close to it.

However, in some cases, people with this kind of blindness can still detect the things in the world that they cannot consciously see. This condition is known as blindsight. In laboratory settings, these individuals can detect visual stimuli that they cannot consciously see. When they are prompted by an experimenter to make a guess about things in front of them, they can use visual processes to predict the thing's location, direction, and color. They cannot consciously see the things they make predictions about. They are unaware of them, blind to their presence. But they can nonetheless "sense" the things in front of them through alternative unimpaired visual pathways. Patients with blindsight have a kind of sixth sense that informs their gray matter about where the thing in front of them is located and what its color is, but the sixth sense does not allow them to consciously see anything.

And with training, blindsighters can learn to detect their environment visually using their eyes, without being prompted by experimenters. For example, one subject was able to navigate around obstacles when walking down a hallway, despite being unable to look at the hallway and see the obstacles.

Some research has suggested that the brain switches from relying on the primary visual cortex for visual navigation to using a rudimentary form of vision that is still used by reptiles. Similarly, there are also reported cases of deaf hearing. Individuals with deaf hearing cannot voluntarily detect, locate, or identify sounds, but when forced to pay attention to sounds, detect them, or determine their location, they are sometimes able to do it.

Both blindsight and deaf hearing sometimes involve minimal awareness (of "something happening" or "something being there"), but subjects with blindsight or deaf hearing lack consciousness of nearly all the things they can detect with their eyes or ears.

Together with our collaborators in the UK and Houston, we tested Lidell for deaf hearing. Lidell was placed in a soundproof darkened room with his hearing aids on and blindfolded to ensure he did not pick up any visual cues from the experimenter. We then played a variety of sounds through the headphones, through either the right or left ear, or both ears. Lidell was asked to determine whether a sound was taking place and whether a sound was played on the right or left, and to identify the sound (for example, a dog barking). This task was split into three parts, beginning with the "with hearing aids condition" which served as a control measure as well as allowing Lidell to become familiar with the sounds he would be asked to discriminate among. At the beginning of each sound trial, one of the three sounds, or no sound, was played in random order repeatedly on a loop. After five seconds, Lidell was tapped on the back of the neck indicating he should raise his right hand to indicate what sound was being played: a boxing move represented "band," a circulating finger represented "ringer," a talking hand movement represented "sonar," and a flat hand represented "no sound." He also was instructed to point his left thumb up or down to indicate whether he heard a sound. This process was repeated with eight seconds of silence between trials. Lidell completed eighty trials, totaling twenty trials for each of the sounds and for silence, each of which occurred with equal probability. Lidell correctly identified which sound was playing or if silence was playing fifty-six times out of a possible eighty, which is significantly above chance, given that there were four possibilities to choose among.

What explains his ability to answer questions about sounds he cannot hear? We believe that he doesn't use the regular ventral-stream perception pathway to process auditory information. The processing of auditory content is in some ways analogous to the processing of visual content. Much like with the visual system, there is segregated processing of auditory information in anatomically and functionally separate ventral and dorsal processing streams. Functionally, the ventral and dorsal auditory streams correspond to auditory stimuli involved in perception and auditory stimuli involved in action guidance, respectively. In hearing for perception, the auditory stimulus is processed in the auditory cortex and is then transmitted directly to working memory in the prefrontal cortex. In hearing for action, the auditory stimulus is processed in the auditory cortex and is then transmitted to the parietal cortices.

It is plausible that Lidell was relying on the dorsal auditory pathway for auditory content discrimination. In all other known cases of deaf hearing, participants sustained damage to previously functioning auditory systems. Lidell's deaf hearing, however, likely was present during early development. At that time, his brain could not rely on auditory information to detect and localize potential action-guiding stimuli, so the particular areas of the auditory dorsal stream that normally process action-guiding auditory information were rendered useless. Because of a diminished capacity to use hearing for action, the dorsal auditory pathway might have reorganized itself in such a way as to process auditory information that he can access in forced-choice paradigms. Dorsal stream representations in the vision for action pathway are not normally associated with conscious experience. If the same is true for dorsal-stream representations in the hearing for action pathway, this would explain the reported lack of conscious experience, despite Lidell's ability to correctly choose among several possibilities for what he heard.

Hearing Lights Blinking

The case of Lidell is similar to those of acquired savants in one crucial respect: In many supermind cases visual imagery or synesthesia appears to be a gateway into the brain's unconscious processes. Like Jason and Derek, Lidell is a synesthete. He hears sounds in response to things he sees in his environment. Lidell probably developed his synesthesia after getting hearing aids at age five. At that age Lidell spent a great deal of time sitting in front of the television set glued to cartoon shows, absorbing everything. He was fascinated by cartoon sound effects: drips, kisses, horns, boings, squeaks, pings, pops, swooshes, and whips. When not spending time in front of the television, he would imagine things in the real world making the same sounds. It soon became habit for him to imagine things making cartoon sounds. Eventually he could no longer undo the connection between things and their sound effects. His synesthesia was born.

Many people wonder what synesthetic sounds, or syn-sounds, sound like. What is it like to "hear" syn-sounds? This is not an easy question to answer. Asking Lidell what it means to "hear" syn-sounds is like him asking you what it means to hear English. Lidell tells us that his syn-sounds are just as loud as normal sounds. But his syn-sounds are not coming through his ears. They are triggered by visual features of reality. Just about anything he sees will have a sound. People can sound like drips, kisses, horns, boings, squeaks, pops, swooshes, whips and so on. Lidell tells us that some people have awful syn-sounds. He once met someone who sounded like a DJ scratching the vinyl with a needle: SCCRRRRRRRRRCCCCCCT! That sound is so nerve-wracking to Lidell that he would do almost anything to avoid being around the person who had the sound. Another synesthete named James Wannerton, who has word-to-taste synesthesia, could probably sympathize. Wannerton once had difficulties visiting his friend and his friend's wife, because the wife's name tasted like vomit.

For Lidell, the sound of the strip on the highway while driving is "plink plink plink plink." The sound of a road sign approaching is "whoooosh." Cars approaching sound like a "whine" and then a "whoosh" when they go away. The sound of a smile is bright and steadily moves to a higher pitch. The sound of a frown goes to a lower pitch, a bass sound. The sounds of people's eyes constantly change during the course of a conversation. The smell of Febreze has a high-pitch twinkling sound. A skid mark in Lidell's shorts from the hamper has a "glaring low-pitch gnarly sound." A dog sounds like a rapid-fire series of clicks. Cats do not have any sounds. "Maybe this is why I like cats," Lidell says.

The sounds of foods completely determine whether Lidell will like them. He tastes food through its sound. Pizza sounds a lot like spaghetti with its spiciness, Lidell explains—it has a sharp pointed sound. Beer with its bitterness has a "flubbing" sound: "brrrrrrrrrrrrrrrrrrr." Sushi has the sound of slurping goo. Carbonated drinks sound like micro-fireworks. Scallops sound like something is breaking. Lidell says that the food with the most horrible sound was a Bolognese loaded with sliced sausages, which sounded like "a train wreck collision of metals." And the loveliest sound: Lidell's homemade German beef rouladen. "The carrots, onions, celery, bacon, pickles—a symphony on your tongue."

Eating in a very noisy environment can interfere with Lidell's taste syn-sound when Lidell is wearing hearing aids that allow him to hear some sounds. When he cannot hear his food, it begins to lose its flavor. He never liked eating in a noisy setting. The food quickly gets bland and he loses his appetite. As Lidell recalls:

> Recently I had a most beautiful and tasty lamb shank. Sadly the restaurant was quickly full of people and all the noise of the chitter chatter was so great that I could no longer hear the taste. That was when I lost all the flavor. Just bland. Turning off my hearing aid would not help since a little ambient sound helps the taste. Russian synesthete Solomon Shereshevsky said the reason soft music is played in restaurants is to make the food taste better. I agree.

Initially it wasn't clear to Lidell that not everybody heard the synesthetic sounds he was hearing. He asked a childhood friend if he could also hear the red lights atop a radio tower flash. The boy took Lidell to be crazy. "You can hear lights blinking? Lights don't have a sound, you idiot." His friend started laughing. "You are nuts. Insane." For several years Lidell thought that perhaps he was hearing things he saw because he was mentally ill. When he got older he briefly feared that he might have paranoid schizophrenia, for auditory hallucination is a common trait of the condition.

Because of this fear, Lidell started to pay constant attention to every aspect of his inner sounds and eventually realized that he was experiencing something no one else could. He had read in some books that the sounds schizophrenics experience often feel as though they are coming from some outside source. They are not experienced as part of the person's agency. But Lidell felt sure that his sounds did not belong to another person or an alien agency. They were "his" sounds.

Faces in the Crowd

People have their unique syn-sounds for Lidell, which helps him recognize them. This "ping sound" makes it possible for Lidell to quickly identify familiar faces in a crowd. The "ping" experiences are the most complex of Lidell's synesthetic experiences. Each one of us categorizes the world in our own idiosyncratic fashion. When looking at faces, we pay attention to different facial features and other characteristics. One of us may focus more on the beard, another on the mole on the nose, yet another may notice the dimples on the cheeks. None of us sees exactly the same face. Normally there are certain cues that enable us to recognize other people even when they shave off the beard or have aged years by the time we see them again.

Though Lidell can see faces without hearing pings, he says he can't recognize a face unless he hears a ping, even if he has seen it before. He remembers an instance when synesthete Pat Duffy stepped out of the elevator at a conference and greeted him by name. He knew her well, that was clear, but it was not clear to him who she was; he was waiting for a ping, but didn't hear one. So he stood there with a blank stare as she talked to him. After several seconds he finally heard a ping and remembered. All he could say was "Pat!"

This appears to be a type of face blindness, also known as prosopagnosia. Face blindness can be extremely debilitating. The phenomenon is particularly well illustrated in *Faces in the Crowd*, a horror-thriller from 2011. The main character survives the attack of a serial killer but the incident leaves her with mild brain damage to the region that processes faces and allows us to recognize them. The main character is taught to recognize people by consciously taking note of details other than their faces, such as their ties, shoes, jackets, or hair color. What is particularly unnerving about the movie is how it uses the fact that a large percentage of us have a mild form of face blindness: We have no problem with faces in general but get confused when faces are similar. The male characters in the movie look so much alike that even the slightest face blindness will leave you feeling as lost as the female protagonist.

The Clinton-Gore illusion illustrates how easy it is to play tricks on our ability to recognize faces. Here "Gore" and "Clinton" have the exact same face. (Reprinted by permission from Macmillan Publishers Ltd: NATURE, "I think I know that face . . . ," Pawan Sinha and Tomaso Poggio, copyright 1996.)

Lidell reports that "ping" sounds counteract his face blindness. When people he knows enter the periphery of his field of vision, he hears a ping. Strangers that he has paid no attention to in the past do not give rise to a ping, unless the strangers for some reason really have caught his attention.

Lidell's pings come from wherever the face is located, even if he is not consciously aware of having seen anyone. He tells us that once in the crowded lobby of the Natural History Museum in London, he suddenly heard a ping coming from the other side of the room. When he walked across the room, he realized it came from fellow synesthete and friend Bryan Alvarez. He was sitting with his head down, reading the museum guide. Lidell tells us about another incident. He was driving north on I-55 going about eighty miles an hour. He casually looked at the cars going south in the other lane. Suddenly he heard the ping. He crossed the median and headed south to track down its source. When he finally caught up, the source of the ping's car was there: his childhood friend, but with a new appearance. His Hell's Angels look had been replaced by a clean, shaven one.

Lidell's "ping" experiences appear to be due to a hypersensitivity in his peripheral vision. If someone were to wave a hand in the periphery of Lidell's visual field, he would hear an accompanying syn-sound. Fluorescent lights, with their rapid flickering, are particularly obnoxious. When he is focusing on a task like finding an error in a thousand lines of programming code, the office lighting can make his job near impossible. In situations like these, syn-sounds can be vampires. The longer the movement, the stronger the sound, and this takes away the energy Lidell needs to focus on the task at hand.

What could be the mechanism underlying Lidell's enhanced peripheral vision, his "sixth sense"? Well, we know that the brain is excellent at accommodating changes and losses. Individuals who experience deafness from an early age sometimes undergo brain rewiring. Some early-deaf individuals, like Lidell, apparently develop superhuman abilities to locate objects in space and to detect motion. Stephen G. Lomber, a professor in the Department of Physiology and

Pharmacology at the Schulich School of Medicine & Dentistry, and his colleagues recently investigated the auditory cortex of congenitally deaf cats. They found that in deaf cats a region of auditory cortex normally involved in locating sounds in the periphery of the visual field had rewired itself to respond to peripheral visual stimulation. Though the input would be different after the reorganization, the output remained the same. Apparently, the auditory cortex can reorganize itself to take visual input instead of auditory input without having to change the kind of information it sends to the rest of the brain and body.

The researchers also looked at what would happen when they inactivated the region in the auditory cortex by cooling it. They found that when it was cooled down, hearing and deaf cats alike would lose their ability to visually detect and locate things occurring in the periphery of their visual field. Visual stimuli that occur in the periphery of the visual field thus seem to be processed in the auditory cortex. The results of the study led the researchers to conclude that in deaf animals the behavioral role of the auditory cortex stays the same, but switches from auditory to visual. In hearing individuals, visual inputs give rise to a mild activation of neurons in the auditory cortex. However, neurons respond much more strongly to sound stimulation. In animals that are profoundly deaf, the neurons progressively become more and more sensitive to visual stimuli.

Lomber's study helps explain Lidell's enhanced peripheral vision. Since he was born profoundly deaf, the neurons in the auditory region of the brain did not receive any significant sound input. So over time they may have become progressively more sensitive to visual stimuli. At some point they likely became hypersensitive to visual input in the periphery of Lidell's visual field, giving him the ability to rapidly detect a face in a crowd.

Christina M. Karns, a postdoctoral research associate in the Brain

Development Lab at the University of Oregon, Eugene, and her colleagues, conducted another study confirming similar results. The research team invented a device that could be worn like headphones during a brain imaging study. Two soundless puffs of air were blown onto the subject's right eyebrow and to the cheek below the right eye. The subjects were simultaneously exposed to a brief flash of light.

The researchers modeled the setup of the study on what is known as the "double flash illusion." In this illusion, which occurs in hearing people, one flash of light combined with several brief sound events causes subjects to experience two or more flashes of light. The deaf subjects who were exposed to one flash of light and two air puffs in Karns's study saw two flashes of light. The study showed that deaf subjects had greater brain activation in the auditory cortex than hearing subjects both in response to the air touch and to the double flash illusion. This indicates that visual information and touch sensations are processed and interact with each other in the auditory cortex of deaf individuals.

The old saying goes that we only use a fraction of our brain's capacity. It turns out that tapping into superhuman ability can involve tapping into areas of the brain normally reserved for other, seemingly unrelated tasks. As we will see in chapter 9, technology is now harnessing the power of neuroplasticity, allowing us to substitute digital information for sensory experiences.

Hearing Music in Color

An anonymous person wrote this letter after hearing a seven-year-old Mozart perform: "I saw and heard how, when he was made to listen in another room, they would give him notes, now high, now low, not only on the piano but on every other imaginable instrument as well, and he came out with the letter of the name of the note in an instant. Indeed, on hearing a bell toll, or a clock or even a pocket watch strike, he was able at the same moment to name the note of the bell or time piece." Mozart had what is called perfect, or absolute, pitch. Recall that perfect pitch is the ability to identify or sing a specific note without any reference point. This isn't just a party trick. It's a fundamentally different way of hearing. When you have perfect pitch, harmonic and melodic analyses require virtually no effort. Perfect pitch enables you to replicate a song on any musical instrument you know how to play by simply hearing it. It's like *understanding* a language versus simply hearing the sounds spoken. It puts you in a position to hear a layer of meaning that no one else can hear.

Perfect pitch is often contrasted with vision. When seeing a yellow coffee mug, you can immediately tell that it is yellow. If color were like pitch, determining the color of a mug would require seeing it as contrasted against, say, an orange and green cup. Likewise, if color were like pitch, it would be difficult to determine that the color of the sky and the color of American mailboxes is the same in terms of hue, viz blue.

The ability to identify hue without a reference point is universal. However, there are limits to our ability to determine the precise color of objects. When our brains compute representations of three-dimensional objects, they rely on illumination across the surface of the object. We don't normally pay much attention to that shading, and we certainly are not usually very good as retaining that kind of information in memory. This is why so few of us are naturally talented at drawing realistically. Those who can retain that kind of information about illumination tend to be skilled painters and drawers. As we saw in Chapter 6, they perform what neuroscientist Allan Snyder calls a "right-brain shift" by turning off the normally automatic interpretation of raw sensory data and focusing on things like illumination patterns. So, there may well be a closer analogy between the visual and the auditory cases than most people realize.

Perfect pitch seems crucial to the success of people like Gloria Lenhoff, the Williams syndrome patient with a very low IQ but extremely sophisticated musical skills. Similar musical gifts have also been witnessed in children. Adorable seven-year-old pianist Elias Phoenix from Florida became an Internet sensation after appearing twice as a guest of Ellen DeGeneres on her US talk show. He had earlier performed in the world famous Carnegie Hall in New York City and placed second in his age group at the American Protégé International Music Talent Competition. Elias's mom taught him and his twin brother, Zion, to play the piano using sheet music printed off the Internet.

There is good reason to think that perfect pitch is something all of us start out with, but lose as we age. Infants preferentially respond to absolute rather than relative pitch differences. This suggests that we most likely are all born with perfect pitch. But for most people, this ability begins to diminish as they start acquiring and interpreting language.

In most languages, including English, tones do not play a role in interpretation. Rather, identifying words by pitch would interfere with language interpretation, as there are huge individual differences in the frequencies of people's intonations in most languages. In countries where people speak tone languages, by contrast, there is almost no difference in the frequencies of people's intonations. What matters in English, unlike in tone languages, then, is not the pitch of sounds (or frequencies) but more holistic sound "images." So, when the brain learns to interpret language as having a meaning that goes beyond the tones of the language, the ability to detect tones without a comparison class is lost. This may be one of the reasons that it is difficult to acquire a second language without retaining some accent after the teen years. The brain becomes less focused on the phonemes of a foreign language and more focused on the meaning of the words.

There are exceptions to the claim that perfect pitch is lost during childhood or puberty. Children who undergo musical training at a young age sometimes retain perfect pitch, even though musical training typically only involves learning relative pitch. Furthermore, a significantly greater number of people who speak tone languages, e.g., Mandarin and Vietnamese, which use relative pitch, have perfect pitch compared to Americans who speak English, which is not a tone language. Likewise, Asians who no longer speak a tone language but who spent their early childhood in Asia are more likely to have perfect pitch. Music psychologist Diana Deutsch conducted a study of 203 students at the University of Southern California Thornton School of Music. The students were separated into four groups: one group of Caucasians who didn't speak a tone language fluently, and three groups of individuals of East Asian descent varying in how fluent they were in a tone language. Each group was further divided depending on whether the students had begun musical training between age two and five, or between age six and nine. Deutsch found that fluent tone language speakers scored over 90 percent on average, compared to less than 30 percent among the Caucasians and less than 40 percent for East Asians who were not fluent in a tone language. The results strongly point to cultural factors rather than genetic factors in developing perfect pitch. Musical training, of course, is needed. Without any kind of musical training, you cannot identify musical notes as particular notes. But musical training is only a necessary ingredient; it is far from sufficient. According to Deutsch, the simplest way for a parent to cultivate perfect pitch in a child outside of teaching them a tone language is to expose them to notes and their names in the first couple of years of life.

Most people believe that the time period for acquiring perfect pitch is brief. If the skill is not retained before age seven, it cannot be regained, according to common wisdom. This belief, however, turns out to be false. A new study, conducted by Takao Hensch, professor of molecular and cellular biology at Harvard University and colleagues, shows that valproic acid can provide people with the ability to learn perfect pitch after that critical period has closed. Valproic acid is an anticonvulsant and mood-stabilizing drug used primarily in the treatment of epilepsy and bipolar disorder, and the prevention of migraine headaches. There is also a strong connection between prenatal exposure to valproic acid and autism in humans. In Hensch's study, participants with no prior musical training were given valproic acid. These participants, along with a control group, were then given exercises for two weeks that were aimed at improving their pitch. The group on the drug performed significantly better at the end than the control group, who were given a placebo.

The researchers believe that valproic acid has the ability to increase the brain's plasticity and possibly return the brain's ability to undergo changes to a "juvenile state," allowing it to learn skills that only children can normally learn. Valproic acid inhibits a chemical known as histone deacetylase, which leads to significant alterations in the expression of multiple genes, and in turn allows for a rewiring of the neurons in the brain; in particular, it may increase white-matter connectivity.

But for those who are not interested in taking brain-altering epilepsy drugs to improve their pitch, there are other options. To understand them, we look to another group in which perfect pitch occurs more often than in the general population: synesthetes. Using multiple senses to process information gives rise to abilities that don't usually develop through typical sensory processing.

Many people who are born with perfect pitch are synesthetes. They "see" musical notes as having a concrete identity and presence, usually in the form of colors and three-dimensional shapes. But being born with this ability isn't the only way to get it. Nonsynesthete children can acquire synesthesia-based perfect pitch, with the right training. The Eguchi Method Perfect Pitch Training System, developed by Kazuko Eguchi, has been successfully used to train children youger than six to associate color and pitch. One way this program differs from other methods of pitch training is by focusing initially on chords instead of single notes. Eguchi says that starting with notes instead of chords leads some children to identify the note by its relative position to another note.

In Eguchi's system, children listen to chords and match them with colored flags without providing an actual pitch name. At home a parent starts by playing, say, the three-note C major chord on a piano, and the child, who cannot see the piano, is asked to raise a red flag. They repeat this exercise with the C chord a few times every day, for a few minutes each time. The parent then adds, say, an F major chord and the child raises a different color flag. Chords are always played in random sequence to prevent the child from identifying any chord by its relation to another. All the white-key chords are associated with a colored flag, then all those with black keys. The child names the chord only by its color. Later, the child calls out the individual notes that make up the chord. For C major, which is C-E-G, or do-mi-so, the child raises the red flag and says, "red, do-mi-so," for example. Eventually, the parent plays the chord and then plays the highest note of the chord separately. The child names the chord and the individual notes, and then, upon hearing the single note, identifies it.

According to Eguchi, perfect pitch is trainable in children because their young brains are plastic enough to generate new, lasting connections, something that older brains cannot do as easily. However, perfect pitch can be taught to adults as well. Although it is true that the adult brain is less plastic than a child's, the adult and the child brains are equally malleable with respect to emotional memory.

By using a variation on the Eguchi Method Perfect Pitch Training System you can train yourself to associate specific musical chords with colors and emotions. This method is similar to other techniques used to create automatic synesthetic associations in nonsynesthetes, and can build upon a person's natural preferences for associating colors and chords. Many people naturally tend to have a loose association between certain colors with music, the way they might associate, say, blue with sadness.

The method is best carried out with two helpers, one to play the piano and one to generate the emotion. On the first day, one helper starts by playing the three-note C major chord on a piano, and the learner, who should not be able to see the piano keys, raises whichever one of the twenty-four colored flags that seems to best fit the chord. The learner names the chord only by its color. It's not time yet to worry about identifying it as a C major chord—at this point, the idea is to build a firm association between a particular chord and a particular color. The second helper has a repertoire of emotion-generating sounds on an iPad or some other device and chooses one that fits the color the learner chooses. If the learner chooses a blue flag for the three-note C major chord, the second helper chooses a sound that normally produces a strong feeling of sadness, such as a crying sound effect, and plays it following the flagging. Then the pianist switches to, say, an F major, and the learner chooses a new flag, for example red. The second helper plays an angry sound. Four chords are played during a total of four sessions in one day. The same four chords, colors, and sound effects are repeated the next day while adding four more, and so on, until all twenty-four basic chords are in play on day six. To prevent the learner from identifying any chord by its relation to another chord, chords should never be played in the same order. During this time frame the learner learns the different chords as red, blue, or green.

Starting on day seven, the learner says the individual notes that make up the chord out loud after lifting the flag and hearing the sound effect. For example, for C major, which is C-E-G, or do-mi-so, the learner raises the red flag, hears the sound effect, and says, "red, do-mi-so." During this time frame the learner comes to associate chords with their respective names. That is, she goes from thinking "this is the blue chord" to thinking "this is the blue chord, which is a C major."

On day thirteen, the piano helper plays the chord, and the learner raises a flag, hears the sound effect, and sounds out the color and the notes. The pianist then takes the highest note and plays it separately. The learner names the single note. Then the pianist takes the middle note and plays it separately. The learner names the single note again. And the same with the lowest note.

On day fourteen, notes are played at random, and the learner names the note. After two weeks people tend to score correctly 80 percent of the time on randomly chosen notes. If synesthesia is taught first, this goes up to about 85 percent.

The Eguchi method is effective, especially in children. The problem with it is that it's too elaborate for someone to realistically be able to do at home—it requires a piano, colored flags, an iPad, and two people who are willing to help you several times a day. Luckily, there is an electronic version of the Eguchi method, but without the emotional input. It's available online at http://youtu.be/nDTRxrxU4W8.

But there are other, more effective ways to foster perfect pitch. Robert Zatorre, a professor of neuroscience at the Montreal Neurological Institute and co-director of the International Laboratory for Brain, Music and Sound, found that people with perfect pitch and people without perfect pitch process musical notes differently. Musical notes are processed in the posterior dorsolateral frontal cortex, which is used when memorizing associations. So, to find out the most effective method for acquiring perfect pitch, we need to look at methods for complicated memory retrieval. Recall the recommended method for learning how to memorize long strings of arbitrary digits, such as the number pi. The key to perfect memory in this case is the narrative. Each digit is associated with an event in an emotionally meaningful narrative. This is easier to do with digits than with musical notes. This is because digits and many two-, three-, and four-digit numbers have a preexisting meaning to us. Seven is the number of dwarfs in Snow White, 12/25 is the first day of Christmas, and 911 is the date of a major terrorist attack. The turn of the century was the year 2000. Perfect pitch consists in associating this sort of meaning with musical notes. Just as you can recognize seven as the number of dwarfs, regardless of whether it is written as "7", "seven," or "vii," so you will need to develop the ability to recognize, say, a D regardless of whether it's a high or low note and regardless of whether the note is played on a piano or a flute or a bell. To do this you need to remember notes in a narrative context. We all remember the famous opening of Beethoven's Fifth Symphony in C minor: G–G–G–E♭ or "Dun dun dun dunnnnn." This is also the Morse code for V or victory. G's association with Beethoven's famous four notes is likely going to be part of the narrative context for G. Beethoven can help you learn how to distinguish between G and E on a site like AbsolutePitchLessons.com, selecting just those two notes. Or consider Pharrell Williams's four-note openings, for example in "Happy." F–F–F–F. Now, add F to the mix when you practice identifying notes. Ideally, you will have a vast number of meaningful events and facts associated with each note. Over time you won't need to recall the narratives to identify the notes. At that point you have got perfect pitch.

CHAPTER 9

Brain Tech

The Future of the Human Brain

What would happen if we could take a pill allowing us to use all the processing power of our brains at once for any task? In the movie *Limitless*, starving writer Eddie Morra (played by Bradley Cooper) gains access to an experimental drug called NZT, which allows him to do just that. At first, NZT functions as the perfect stimulant, endowing Morra with the almost limitless ability to gain a bird's-eye view on every aspect of his life. He suddenly becomes inspired to finish the last ninety pages of his book. He borrows a hundred thousand dollars from a loan shark and swiftly makes a two-million-dollar return, catching the attention of a business tycoon, who seeks his advice on a massive corporate merger. He even rekindles his relationship with an ex-girlfriend, who had been out of his league only a few weeks prior. But, like many other things, NZT is too good to be true. People start realizing that Morra is in possession of what is essentially the entire stash of NZT. Loan sharks come after him for a piece of the pie, the tycoon tries to extort him, and, to add insult to injury, the drug's rebound effect is deadly. Morra turns out to be not so limitless after all.

We admit that, despite the negatives, we'd love to get our hands on some NZT. Who wouldn't risk a few dangerous side effects to fulfill every dream of success one ever imagined? The problem is that the drug doesn't exist. A super pill has yet to be invented, and it's a long way off. But the limits of human potential are indeed expanding through other (real) ways to enhance cognition.

We've seen throughout this book that brains can learn to function in unconventional ways in many different circumstances. Novel talent might appear swiftly after a no-nonsense head injury. Such is the case with Denver teen lacrosse player Lachlan Connors. After two concussions that forced him to quit playing, he developed an interest in music. After virtually no practice he learned to play a variety of instruments, thirteen to be exact.

Novel talent, as we have seen, might also emerge unhurriedly though earnest exertion. Folks like Mark Nissen and Daniel Tammet fall into this category. But as our technology improves, we are beginning to discover even more ways of bringing the innate talents of ordinary people into the light of day.

Certain types of vitamins, supplements, and medications have the potential to optimize our brain function. Most of these so-called nootropic drugs work on the attention centers of the brain. So they have a rather specific effect: They enhance our working memory and impulse control. The results of a 2008 study published in *Nature* indicate that more than 20 percent of scientists have used brain-enhancing drugs at some time. Could it be that these scientists are onto something? Possibly. In order to make that sort of judgment, we'd have to know what they were taking as well as how productive they are as individuals. Are they publishing papers, doing groundbreaking research, and coming up with original ideas? We'd also have to consider whether the competitive edge provided through the use of drugs exceeds the harm from potential risks.

Some argue that performance-enhancing drug use should not be advocated under any circumstances. They point to the overmedication of today's children or the scandals generated by the likes of Lance Armstrong and Alex Rodriguez to demonstrate that the use of nootropic drugs will lead to complete dependence and unfair advantages. They believe that taking medication to enhance our brains is cheating, and that increasing our performance is more properly achieved through hard work. But while overmedication is a serious problem, taking a "brain mint," so to speak, is not in itself a troubling notion.

To see why not, just consider how much coffee the average American drinks every day. We hate to break it to you, but the only significant difference between caffeine and amphetamine-based drugs like Adderall is that caffeine is found in many natural food products. Both drugs stimulate the central nervous system, leading to the "speedy" feeling a user feels a short time after taking them. Both give the user significantly increased alertness. Both cause increases in blood pressure as well as urine production. Both have a pretty significant tolerance effect, requiring that more and more of the drug is regularly consumed to experience the positive effects. And both have rebound effects. This means that cognitive function decreases below baseline when the user stops taking it. In fact, caffeine's pharmacological properties fulfill the exact same criteria that led to Adderall's strict regulation. Much of the reason that caffeine has evaded similar heavy regulation by the federal government is that it would be too hard to remove from the food supply. You can thank the lobbyists for that. If it weren't for them, your coffee would be classified in the same category as Adderall.

There are some over-the-counter supplements that provide great performance-enhancing effects without all the side effects of regulated narcotics like Adderall. Adrafinil is one such drug. When ingested, adrafinil breaks down into modafinil, the same compound that makes up the active ingredient of a narcolepsy drug that the FDA approved in 1998. But it's not only good for combating random sleeping spells. The drug turns out to have very beneficial effects on memory, concentration, and wakefulness in general. Given that it's not currently regulated by the Drug Enforcement Administration (DEA), adrafinil is available to purchase and possess without a prescription.

Like modafinil, adrafinil affects the adrenergic receptors in the brain, causing the release of the neurotransmitter norepinephrine, combating fatigue and enhancing concentration. The drug is also thought to increase levels of glutamate while suppressing levels of GABA, the brain's main excitatory and inhibitory neurotransmitters, respectively. Recent studies show that this effect can help speed up thought processes and focus so you can quickly complete the tasks on your plate. Keep in mind, though, that since this is a drug, there's potential for interaction with other drugs. So make sure you check with your doctor *and* pharmacist before ordering a bottle.

Another drug with some great potential for brain enhancement is psilocybin, the magic ingredient in psychedelic mushrooms. They appear to have the same potential that a hard hit on the head can have. Jason and Derek became extremely attuned to artistic activities after having their brains shaken up. Participants in recent psilocybin studies have developed similar creative or mathematical abilities. They still had these abilities after eighteen months.

How could psilocybin affect the brain in the same way as banging your head? Psilocybin is a serotonin agonist—it imitates serotonin in the brain. Admittedly, serotonin is not known for sparking creativity. In fact, there is at least an anecdotal connection between depression (which is accompanied by low serotonin levels) and creativity. Depressed people who are highly creative sometimes allege that they cannot take antidepressants because it will kill their creativity. However, typical antidepressants work very differently from psilocybin. Antidepressants keep serotonin levels steady and stable over a long time interval. A one-time moderate to high dose of psilocybin, on the other hand, acts like a jumper cable on a car with a dead battery. It kicks the whole brain into on-mode by acting in many different areas of the brain. The serotonin receptors that psilocybin acts on are all over the brain's cortex and throughout some of the subcortical structure, which is the seat of emotion, memory, motivation, desire, and instinct.

Psilocybin may be a seriously needed performance enhancer. It is known to increase people's openness to new kinds of activities and new ways of adding up old information. This appears to be a necessary ingredient in scientific discovery and creativity in general. When you look at case reports of people who have had breakthroughs, a good number of these breakthroughs didn't happen while the scientists were busy at their desks, doing the work they were hired to do.

It's been rumored that the double helix structure of DNA occurred to geneticist and neuroscientist Francis Crick while he was tripping on LSD, which is chemically similar to psilocybin and affects the brain in the same way. He won the Nobel Prize for his discovery.

It is widely believed that Kary Mullis hit on the idea behind polymerase chain reaction (PCR), a now widely used technique for amplifying a single piece of DNA by a factor of 100 billion, the same way. The technique is essential to DNA testing, detection of infectious and hereditary diseases, and DNA sequencing. When Mullis found the solution in 1983, he was a low-level tech working for Cetus Corporation in Emeryville, California. The story is that he thought of the PCR while on LSD, cruising along the Pacific Coast Highway one night in his car. The whole thing at once presented itself to him. When the news got out everyone was smacking their foreheads because it was such an obvious solution to a problem molecular biologists had tried to solve for years. Like Crick, he won the Nobel Prize for his discovery.

The idea of giving someone a drug to improve performance and creativity is radical. We don't like it when a cyclist or a bodybuilder takes steroids. But we don't seem to mind when musicians smoke a bunch of weed or alcoholic writers drink a lot of booze to produce things that entertain us. (Would many of the Beatles' legendary songs exist without experimentation with acid? It's no secret that the lyrics of many of the pop legends' famous tracks were inspired by LSD, including "Lucy in the Sky with Diamonds," "I Am the Walrus," "Tomorrow Never Knows," and "What's the New Mary Jane." The Beatles' creating during a hallucinogenic trip is hardly a rare case of acid-driven creation, invention, or discovery.) What if there is a way to use compounds to promote optimal functioning and creativity? As a society we would have to decide whether we are okay with this way of producing art and doing science.

Drugs are only one way to enhance cognitive performance. Research has shown that altering the balance of activity between the right and left hemispheres of the neurotypical brain can strengthen abilities that tend to be enhanced in savants, such as the ability to solve difficult puzzles, quickly count large numbers of objects, and maintain attention to detail.

A Balanced Brain

Electric and magnetic stimulation have the capability to induce savant-like skills in ordinary folks by inhibiting left-brain control centers, thereby leaving room for right-brain creativity. These findings on ordinary folks confirm that part of what prevents us from reaching the level of savants is that our executive areas suppress other brain regions that have been associated with originality and creativity.

In one study, Allan Snyder, director of the Centre for the Mind at the University of Sydney and the Australian National University, showed that drawings of images recalled from memory significantly improved in detail with the use of transcranial magnetic stimulation (TMS), a noninvasive method of interfering with the neurons of the brain. It works by inducing weak electric currents by means of a rapidly changing magnetic field. Participants were given two tasks that involved drawing. In the first task, participants chose between drawing a dog or a horse. In a second drawing task, participants were shown an image of a female face for thirty seconds and then given one minute to draw the image from memory. The participants were then given TMS stimulation for both ten and fifteen minutes. The images that the participants drew after TMS were remarkably improved in detail. The effect appeared to last even forty-five minutes after the stimulation was received.

Journalist Lawrence Osborne went to Australia to try TMS out for himself. Despite having no prior drawing abilities, he suddenly got the hang of drawing cats. He wrote about his experiences for the *New York Times*:

> Two minutes after I started the first drawing, I was instructed to try again. After another two minutes, I tried a third cat, and then in due course a fourth. Then the experiment was over, and the electrodes were removed. I looked down at my work. The first felines were boxy and stiffly unconvincing. But after I had been subjected to about 10 minutes of transcranial magnetic stimulation, their tails had grown more vibrant, more nervous; their faces were personable and convincing. They were even beginning to wear clever expressions.

1. PRACTICE   2. BEFORE   3. DURING   4. AFTER

These illustrations came from a different session.

Another study showed that inhibiting activity in certain areas of the brain can facilitate solving geometric puzzles. One particularly difficult geometric problem is called the nine-dots puzzle. The goal is to connect nine dots arranged three-by-three using only four lines without lifting the pencil from the paper or retracing the line. The nine-dots puzzle is so difficult that most studies report that almost all participants fail to solve it. And they fail to solve it despite being given hints or a long time to solve it, even after a hundred attempts. The only way to beat the puzzle is to extend some of your pencil lines beyond the imaginary boundaries formed by the dots. It's so difficult because it requires that the participant look "outside the box."

Snyder used transcranial direct current stimulation (DCS) to alter the activity in both the left and right hemispheres. DCS works by passing a direct electric current through the brain, which either decreases or increases the excitability of neurons in a region of interest. Snyder placed the electrodes on either side of the skull, passing the current through the whole brain. This causes each hemisphere to be affected in opposite ways—the excitability of one hemisphere is increased while the excitability of the other is decreased. Before stimulation, participants were unable to solve the nine-dots puzzle. But after DCS was applied for only ten minutes, specifically to decrease activity in the left hemisphere and increase activity in the right hemisphere, something amazing happened—40 percent of subjects were able to solve the puzzle.

Researchers believe that tasks like drawing or solving the nine-dots puzzle are notoriously difficult because our brains are structured in such a way as to limit creativity. Snyder argues that the brain is designed to use concepts from our past experience when interpreting the world. Such a limitation may have an evolutionary benefit: The brain can become more efficient at processing the types of information it regularly comes across in the environment. Survival in the wild requires that one recognize a threat as quickly as possible in order to avoid harm. Being overactive at grouping discrete parts in order to overestimate rather than underestimate environmental threats is no doubt evolutionarily advantageous. But this overactive grouping, while evolutionarily beneficial, limits our ability to process much more abstract information. When we view the nine-dots puzzle, we see it not merely as a bunch of random dots (which it is), but as an organized figure with strict boundaries. This is not something we have conscious control over. Our minds simply limit how we can approach such a problem.

DCS is not focused like TMS because the former device passes a current through one side of the brain to the other while the latter device emits a concentrated magnetic field. In these studies, Snyder inhibited activity of the left hemisphere while facilitating activity of the right hemisphere. This mimics the pattern of brain activity seen in many savants. Inhibiting the left hemisphere is thought to remove the predisposition to interpret random elements in meaningful ways. Interestingly, Snyder notes that one person who was excluded from the study was able to solve the nine-dots puzzle without DCS. At the initial interview, the person reported having sustained a head injury at about ten years of age and was excluded to ensure that all participants had similar backgrounds. But the man was interested in seeing how the experiment worked, so Snyder tested him anyway. It is plausible that the head injury affected the man's brain in a way similar to that of DCS, allowing him to solve the puzzle.

How can TMS and DCS make innate talents emerge? These devices really just add noise to the brain by strengthening neuron signals in small doses. When applied at large frequencies this confuses the neurons and they stop doing their job. Snyder's lab used TMS and DCS to inhibit substantial areas of the brain, namely those areas of the left hemisphere that are responsible for rational decision-making and self-control (the Freudian or Nietzschean superego). Inhibiting these areas allows other brain regions in the right hemisphere to step up and show off their X-factor. It's a bit like removing the bullies from the playground to give the more sensitive kids a chance to shine.

But are there any cognitive tradeoffs? Inhibiting large areas of the brain presumably can affect areas that are crucial for cognitive function. Snyder tried to control for this possibility too. In the DCS study, participants who were stimulated were asked to solve a series of algebraic problems before attempting to solve the nine-dots puzzle. DCS stimulation did not appear to affect participants' ability to solve the math equations. This finding is consistent with other studies showing that inhibiting the dominant motor area and exciting the nondominant motor area improves the motor performance of the nondominant hand while not affecting the motor performance of the dominant hand. These exciting technologies thus appear to be new ways to increase brain performance in at least some sorts of difficult tasks. Thus far, researchers have yet to identify any downsides of using TMS to enhance brainpower.

Researchers are also experimenting with lasers to obtain the same results as shown in the TMS and DCS studies. Sounds like sci-fi, right? This technology has been around for some time, however. Transcranial laser stimulation (TLS) works much like magnetic stimulation, except it uses lasers to affect neurons deep inside the brain. University of Texas at Austin psychologists Douglas W. Barrett and Francisco Gonzalez-Lima have shown that TLS could improve cognitive performance as well as general mood. Using lasers offers some potential benefits over the magnetic method. Magnetic methods of brain stimulation use bulky equipment and draw a lot of power. But laser technology requires a fraction of the space and operating power to potentially get the same effect. So for portable devices, lasers might be the way to go.

Time for an Upgrade

Other novel technology is leveraging the power of neuroplasticity. Recall from chapter 3 that neuroplasticity refers to changes in the brain's neural pathways and synapses resulting from changes in the information processing demands placed on it. One of the hottest studied topics right now involves sensory substitution, the ability to feed one type of sensory information into another sensory channel. Think sound through vision, or taste through touch. Sensory substitution works because the brain doesn't care where the information is coming from; it just cares what kind of information it is.

The concept was first introduced by neuroscientist Paul Bach-y-Rita as a way to substitute touch for vision in blind individuals. The method was originally based on the fact that many sensory disabilities are due to an injury to a sensory organ rather than damage to the sensory processing regions of the brain. So, if under some tragic circumstances you happened to lose function of an eye or ear, you wouldn't necessarily lose the ability to see or hear. If the brain regions that process signals coming from your eyes and ears are still intact, you could still push the signal in through an intact sensory modality. We saw this in a previous chapter that covered human echolocation. Not only is spatial information transmitted by sound processed in part by the visual area of the brain, self-reports suggest that these individuals have genuine visual experiences as a result.

You can see this in a brain scanner. Studies have shown that when blind people are fed two-dimensional visual information through touch sensors, they start to process the touch signal in the visual cortex, the region responsible for vision perception. You can also substitute sensory experience in one area for sensory experience in another. But it doesn't happen immediately. At first, the sensory experience is clouded by noise. Visual information being transmitted tactilely to the skin just feels like a random series of burps and bops. But eventually the brain starts making sense of it all.

This is the idea behind a new type of hearing device being developed as an alternative to the cochlear implant. Cochlear implants are prescribed for individuals whose hearing loss is too severe to be reversed through use of a traditional hearing aid. Sound information picked up by a tiny microphone is passed through a microprocessor, which converts this information into an electrical signal. This signal is passed to the brain via tiny electrodes implanted in the auditory nervous system in the ear. But cochlear implants have some significant drawbacks. Although they allow one to carry on a conversation or pick up potential environmental hazards, the sound experience isn't quite natural. They also require invasive surgery, which makes them incredibly expensive—upwards of fifty thousand dollars.

Neuroscientists David Eagleman and Scott Novich from Baylor College of Medicine are harnessing the power of sensory substitution to create a low-cost alternative to the prohibitively expensive cochlear implant: the Versatile Extra Sensory Transducer, or VEST. The goal of the VEST is to provide deaf individuals with the ability to understand speech without having to read lips or resort to other, less natural methods of communicating.

The VEST works by playing a series of tactile patterns that correspond with different environmental sounds. Sounds are fed into an Android tablet, which converts the information into vibrational patterns. These patterns are instantly transmitted via Bluetooth to a receiver built into the breathable low-profile bodysuit securing a series of vibratory motors arranged in a matrix along the front and back of the torso. These motors vibrate in particular patterns based on the different signals fed in by the tablet.

At first, wearing the VEST is more like getting a massage than taking part in a conversation, for the vibrational patterns seem completely arbitrary. But over time, the wearer starts to pick up on the subtle differences between the patterns. After a period of training lasting a few months, hearing through the vest becomes second nature. Eagleman and Novich think that assistive devices like the vibratory vest are only the tip of the iceberg. In fact, sensory substitution may provide us the ability to enhance our senses way beyond our present understanding of sensory experience. Consumers are also excited about the VEST's potential, as evidenced by Eagleman and Novich's recent success on Kickstarter.

Sensory experience is limited by the hardware you've been given. For example, the human eye is designed to perceive light in the visible spectrum—we don't typically see ultraviolet light like bees do. But that doesn't mean our brains aren't capable of it. In fact, individuals who have had the lens of the eye removed can do just that. Some of the limitations to our hardware are more complex. It's likely that humans will never be born with the olfactory abilities of a bloodhound, for our noses don't have near their number of sensory receptors. These limitations placed on sensory experience are what shape our world experience.

The German philosopher Edmund Husserl called this the "umwelt," often translated as "self-centered world." Every creature has a different sensory experience. Thus while many organisms share the same perceptual environment, they all have a different umwelt. The butterfly might take the experience of seeing ultraviolent light for granted, but couldn't imagine what it would be like to taste a steak, just as many of us could never imagine what it would be like to echolocate. But all of these experiences are merely bites taken out of the same environment.

Sensory substitution provides a way of expanding our umwelt. Instead of merely substituting sight for sound or touch for taste, why not try to port other types of informational signals into our cognitive systems? Imagine being able to predict when it's going to rain by "feeling" the weather patterns in your environment. Or being able to feel the stock market, allowing you to make more intuitive trades. Eagleman and Novich believe that devices like the VEST can accomplish this sort of sensory expansion. At the 2013 Future of Being Human event in San Francisco, they demonstrated one example, dubbing it the "Tweety-sense." During the daylong event the two silently recorded every tweet tagged at the event and translated it in a pattern that you could feel on the vest. Wearing the vest could let you tap into the collective consciousness of thousands of conference-goers.

Using a vest is one of several ways to substitute in sensory experience. Other labs are actively researching devices that utilize static electricity applied to regions like the feet, back, or tongue. It's also possible to implant devices much like cochlear implants. But these methods require substantial engineering before they're ready for prime time. Researchers will need to establish their safety and efficacy before the FDA will approve any of these devices for human use.

The Deathmatch

The ultimate goal of upgrading the brain is to enhance the human experience. We don't just want our brains to be better at performing certain tasks; we want our lives to be better because of it. We want to be able to spend more time with our children because enhancement allows us to spend less time working at the office. We want to see the world in new ways that lead to discoveries benefitting all of mankind. And we don't want to be strapped to a chair like in *The Matrix*—we want this whole thing to occur seamlessly. So just how good can it get? Are there limits to the enhancements that technology can provide to our brains? Will there ever be a day where we can just plug a computer module straight into them?

Given that the brain is a biological entity, finding the answers to these questions is much harder than it might first appear. Interfacing two devices requires that they're able to talk to one another. Think cats and dogs: These two animals often don't get along because they operate in very different ways. Dogs are obedient and loyal. They jump up to greet you at the door. They play games like fetch. With a few exceptions, cats don't do any of these things. It explains why many dogs like children while many cats don't: Dog behavior is much more childlike than that of cats. Dogs and children don't speak the same language as cats, so to speak. So they often don't get along.

The same sort of interfacing problem emerges when you need to connect two electronic devices: They often don't speak the same digital language. Since there often are many ways to do the same thing, competing companies implement their own methods that are most cost-effective to them. So to get them to interface, you have to find some way to get them to talk. The key to this, in digital devices, is called a driver. Your computer's operating system uses drivers any time it's trying to communicate with peripheral devices, such as your printer, Ethernet card, or monitor. The driver acts to translate the language of the operating system into the sort of language a peripheral device understands. This method allows for the free passage of information from one device to the other so that they can communicate.

Brain-computer interfaces would have to use a driver as well. In that case, the sorts of widgets that enhance the brain would function as peripheral devices. A driver would transform the information being output from the brain into the type of signal the digital device could understand, and translate the device's feedback into information the brain could understand.

One such driver system is being developed to allow for the measurement of relaxation. Joshua Jackson and colleagues from Baylor College of Medicine are working on a game called *Meditation Deathmatch*, in which contestants compete to see who can relax the hardest. For each round, two contestants face off to see who can maintain the deepest state of meditation all while an extremely distracting MC, lights, sounds, and even static shocks antagonize players back to a state of awareness. Relaxation is measured via wireless EEG headsets that record the different electrical signals from the brain corresponding with its activity level. The game relies on a driver that translates the raw brain signals into a single numeric measurement of relaxation that can be represented by a meter. The meter tells the judges and the audience just how relaxed each contestant is as the MC berates each contestant back into the waking state.

The same David Eagleman working on the VEST is developing a treatment for drug addiction that leverages a similar type of interface system. Addiction is the result of an imbalance between two networks in the brain that are responsible for craving and impulse control. One network craves the drug, while another network is responsible for suppressing that craving. Beating addiction requires that the impulse control network is strong enough to keep the craving in check. Ricky Savjani, Eagleman, and colleagues believe that it's possible to strengthen the impulse control network through practice. In other words, overcoming addiction is a bit like working out your muscles at a gym. So how do we exercise our brains? Real-time neurofeedback.

The treatment works like this: Crack addicts experiencing a craving sit in an fMRI scanner while a computer records the activation of different brain regions. A computer algorithm calculates the ratio of the activity in the craving network to activity in the impulse control network and converts it to a visual representation, which the addict views on screen. The addict is tasked with trying to decrease the activity in the craving network while increasing the activity in the impulse control network. They can do this in myriad ways, from focusing on some other complex task like evaluating math equations to imagining themselves in a relaxing atmosphere like the beach. The goal is for addicts to start picking up on the particular mental tricks that bring their craving levels down. Over time, the impulse control network becomes strong enough that the addict doesn't need to give it much thought. So much for AA.

Real-time feedback is theoretically promising, but Savjani is careful to note that this type of treatment is still experimental, so it will be some time before you can walk into your local neurofeedback addiction treatment center. One potential difficulty in making this sort of treatment successful is that it requires that the addict is dedicated to overcoming their addiction. But the science is there. Similar studies show this type of neurofeedback can help participants successfully manage pain, regulate emotion, and improve working memory. It's not a far jump to think it will work for addiction too. As brain interfaces improve, it's likely we will see the development of a whole new class of devices that help us manage our own minds.

Measuring Thought

One hot topic among brain scientists right now is the notion of synthetic telepathy, a method for direct communication between the brain and an external device. The benefit is that such communication will not require any hand movements or voice activation. Research on synthetic telepathy started in the 1970s at UCLA. From its conception, the research was aimed at developing neuroprosthetics that can control external electronic devices. The initial research was done on animals, starting with rodents and slowly moving up to primates. In 2008, researchers at the University of Pittsburgh Medical Center succeeded in training a monkey with a neuroprosthetic to operate a robotic arm via thought alone.

For sufferers of locked-in syndrome, a horrifying phenomenon in which an individual loses the ability to move despite being fully aware, the success of synthetic, or artificial, telepathy would be more than a scientific breakthrough. It would give a new chance at life for people who cannot move, talk, or eat, despite being fully conscious. The first neuroprosthetic devices were implanted in human brains in the mid- to late 1990s. In 1997 Emory University researchers Philip Kennedy and Roy Bakay started working with Johnny Ray, who suffered from locked-in syndrome following a brain stem stroke that same year. Ray had a neuroprosthetic successfully implanted in 1998. Over the next four years he learned to control a computer cursor via thought alone. In 2005 tetraplegic Matt Nagle received a neural implant and became the first human to control an artificial hand by thinking about moving his hand. Subsequently he also learned to control a computer cursor, a light switch, and a television set.

To ensure competitiveness in twenty-first-century combat, the US military started aggressively exploring synthetic telepathy in 2008. Military grant money is aimed at researching the possibility of composing and sending e-mails, texts, and voice messages using thought alone. This could be beneficial for use by undercover agents, prisoners of war, and frontline soldiers alike.

How is thought-controlled messaging even possible? The current technique is based on a device that is used to read brain signals, called electroencephalography, or EEG. EEG is the recording of electrical signals, or brain waves, along your scalp. The technique is primarily used to measure deviations in standard brain wave fluctuations and can be used to determine whether a patient has a seizure, is in a coma, or is brain dead. The normal woken brain has brain activity that fluctuates between 8 and 100 Hz. An alert and active brain will tend to have neural oscillations, roughly, in the 40-Hz range in at least some parts of the brain. These brain waves are also known as gamma waves. Alpha waves—oscillations in the 8–12 Hz frequency range—and beta waves—oscillations in the 12–30 Hz range—become more prominent when you are inactive, for example, when passively watching television. Brain-dead people and coma patients can have oscillations that approach zero. And in seizure patients, the brain oscillates faster and more regions of the brain vacillate in the same frequency range. In a grand mal seizure, large areas of the brain flicker in synchrony.

Due to the risks involved with experimental brain implantation, much of the research has been done on monkeys trained to complete a specific task, such as moving a hand to a specific location indicated on a screen, while their neural activity is recorded. Information about the task along with neural activity is integrated to estimate the intentions for movement represented by the neural activity. Over the course of training, the computer interfacing the brain and prosthetic device becomes very good at translating brain activity into different intentional actions.

Like other brain scanning devices, such as PET and fMRI, EEG can read which parts of your brain are most (or least) active by measuring how fast neurons in different regions oscillate. So if we know that a particular part of your brain oscillates in the gamma range when you think "Help. The enemy caught me," but oscillates in the alpha range when you think, "Everything is cool here," we can program a computer to translate these signals into a message.

While we already have commercial EEG games, such as Mindball, that allow gamers to manipulate objects by thought alone, EEG messaging may not be just around the corner for the majority of us. EEG is simply not very sensitive when used on the outside of the scalp. In Mindball the EEG device measures which of the players have neural oscillations in the lowest frequency range. This player's ball then moves across a table. Presumably the one who falls asleep or consumed the most Valium wins.

One of the military-funded researchers, Mike D'Zmura at University of California, Irvine, thinks that while much of the funded research will be utilized by the military, the research will result in the development of commercial products as soon as the technology is there to support it. Twenty-five years ago the thought of sitting on opposite sides of the globe communicating face to face, as we do with Skype, was at best a sci-fi comedy. Now we use it on a daily basis.

One of the main worries people have about thought-controlled text and e-mail messaging is that they will have no privacy left. Dare think a single thought about the woman having dinner at the next table and your spouse will slap your face and leave. But, of course, this is not how these devices will ultimately work. People had the same concerns about Skype-like phone calls.

"Oh no, what will happen if I just came out of the shower and the phone rings," your grandma would ask, looking seriously worried. "Well, the device would have an on-off switch," you would calmly reply, but the worried look would not go away. People apparently didn't trust themselves enough to even think it would be possible not to pick up the phone naked on their way out of the shower.

But that was then. Today hardly anyone picks up the phone. And there probably are very few people who have accidentally accepted a Skype call naked. In ten or fifteen years when thought-controlled texting and e-mailing might be commercially available, it is going to be just as unlikely that your most naughty, rude, or immoral thoughts are going to be transferred by sheer accident to the salesperson at the liquor store or your spouse during your anniversary dinner. In fact, you probably will be more concerned about your kids thinking too much to each other at the dinner table instead of eating their vegetables.

Another, more pressing worry is that controlling your own thoughts is not exactly as easy as controlling your mouth. Consider what happens when you are told not to think about something, like a white bear. How successfully could you follow the instruction? Hardly at all. While such a lack of control is of minor impact right now, the inability to control one's thoughts could end in disaster should the wrong person become aware of what's going on inside your head. Furthermore, what about future data thieves who find a way to read your brain waves without you even knowing, gathering secret information like PINs and passwords? This early in the game, it's hard to know what sorts of issues might crop up as the technology evolves. But the hope is that developers will take these worries seriously enough to curb them before they become a big issue.

Upload to Upgrade

In the near future, we're very likely to see some extraordinary new devices facilitating new ways of interacting with the world. But can we go further than merely bolting a bunch of peripherals to our bodies

like Robocop? Are we forever limited by the biological wetware en-
dowed to us through evolution? Some scientists strongly disagree, believing instead that it's only a matter of time before we can upload our
minds to the cloud. And once that becomes a possibility, they say,
there's no telling what limit to cognition there might be, if any.

The scenario might go something like this: Approaching your el-
der years, you realize that your time on Earth is limited. So you sign
an agreement with a brain bank, a company that—upon your death—
moves your mind from your brain to a computer. Although you're no
longer the same flesh and blood, you continue to live your life in virtual reality for as long as you wish. Or at least as long as the simulation
keeps running.

Such a scenario is not necessarily out of the realm of possibility.
We know that the human brain and computers can perform a lot of
the same functions, like addition and subtraction or estimating the
location of an object in space. The question remains as to whether the
brain operates similarly enough to computers that the same sort of
hardware that drives a computer's operation could do so for the mind.

Whether the brain speaks computer language has been a century-
long debate in the philosophy of mind as well as computer science.
Some individuals believe that the brain does in fact operate like a
computer. But what does it even mean to "operate like a computer"?
In the broadest sense, a computer is a device that manipulates information over a series of steps. Considering that definition, it's hard to
see how the brain could possibly *not* be a computer. After all, we
know the brain performs lightning-fast calculations. And it can apply
algorithms to speed these calculations up, just like computers do. But
the worry, some point out, is that the nature of computation cannot
explain our conscious experiences. Since we aren't ready to abandon
the belief that we're conscious, we have to abandon the theory that
our brains are computers.

Think of it this way: Everything your computer does, from checking e-mail to replaying the rich soprano sounds of Anna Netrebko singing *La Traviata* involves moving electricity around a circuit board. There's nothing more going on. If a brain operates in a similar fashion, the question is why we're conscious while the computer is not. How come computers don't experience the beautiful music they're playing? Is it just by virtue of running a different program that we are lucky enough to be aware, unlike those other, lowly computational devices? There has to be something special about us.

We don't yet know the solution to this problem. The reason is that we still don't have a clear idea of why we're conscious. Philosopher David Chalmers is famous for fleshing out the problem in his doctoral dissertation. Chalmers distinguishes between two problems of consciousness: The "easy problem" is figuring out what combination of physical factors gives rise to consciousness. Given sufficient time and resources, it's likely that scientists will figure out how to build a conscious brain. That it's been labeled the easy problem doesn't mean it's a piece of cake: It's a long way before we figure out how to build a brain from scratch, molecule by molecule. But there is a much harder problem that we may never find a solution to. Here's how Chalmers describes it:

> It is undeniable that some organisms are subjects of experience. But the question of how it is that these systems are subjects of experience is perplexing. Why is it that when our cognitive systems engage in visual and auditory information processing, we have visual or auditory experience: the quality of deep blue, the sensation of middle C? How can we explain why there is something it is like to entertain a mental image, or to experience an emotion? It is widely agreed that experience arises from a physical basis, but we have no good

explanation of why and how it so arises. Why should physical processing give rise to a rich inner life at all? It seems objectively unreasonable that it should, and yet it does.

The "hard problem" is figuring out what consciousness is and why it exists at all. Something about our universe is such that when you put the right stuff together in the right way, it all becomes one conscious unit. Suddenly, we see the emergence of an "I." Faced with this problem, some turn to God. The only way to explain conscious personal experience is that a creator gave us life, they say. However, postulating an omnipotent leader is but one way to explain why our world is what it is. It is possible, in fact, that the nature of the physical world has consciousness at its roots. And thus it isn't out of the realm of possibility that a mind could run on a computer, given the right sort of physical structure.

That opens the door to conscious machines. This possibility has laid the foundation for a hotly researched area called artificial general intelligence (AGI), also known as strong AI. The notion of AGI gets its roots in science fiction, perhaps best depicted by HAL in the 1968 American-British film *2001: A Space Odyssey*. Proponents of AGI believe that at some point in the future, computers will be able to perform any mental task a human being can. Although scientists do disagree about what, exactly, is required for a computer to be generally intelligent, there is some agreement about what it should be able to do at a minimum. For example, it should be able to reason, plan and apply strategy, learn, and communicate in a natural language. Many scientists think that those abilities rely on self-awareness, so consciousness plays a significant role here.

Some futurists think we should be very wary of constructing generally intelligent machines. They warn that technological advances will likely lead to the intelligence level of computers exceeding the intelligence levels of human beings, a point in time called the singularity. The term was coined by John von Neumann who, in a conversation with fellow mathematician Stanislaw Ulam, described a point at which human life, as we'know it, would cease to exist. Proponents of the singularity believe the future holds an explosion of artificial intelligence at which point super-intelligent machines start designing themselves, building more and more intelligent machines until machines are able to vastly outthink human beings.

Some futurists like Ray Kurzweil think the singularity can happen as early as 2045. We don't know what life after the singularity would be like, mainly because we don't know what kinds of actions these intelligent machines might take. But if they're anything like us humans, we shouldn't expect their intentions to be in our best interests. What would keep these machines from exploiting us for their benefit? What if they choose to keep us as pets? Not to worry, you think. Can't we restrict these devices so they can't harm us? For example, wouldn't it be possible to program these computers so they'd have to follow our instructions? It's not likely. A machine sufficiently intelligent would be able to predict our actions and secure itself from outside attack. And even if we could restrict their behavior, all it takes is for a lone engineer to build one unrestricted system. At this point, we're doomed.

As far-fetched as it sounds, the singularity is a real possibility. It's become particularly obvious in the last decade. Computers control everything from our communication devices to our lights and even our water supply. Just recently a Hong Kong–based venture capital firm, Deep Knowledge Ventures, appointed an intelligent robot to its board. The goal is for the device to become better at predicting movement in the market over time, eventually gaining a full vote on the board. It's easy to see how this sort of intelligent device could use its power for nefarious purposes.

By the time we make it to this level of AGI, we're likely to understand enough about the brain that uploading minds to machines shouldn't be hard. In fact, we might find it's more efficient to upload minds than it is to enhance or repair the brains that drive them. At this point we don't need to worry about an interfacing problem because brains and computers might speak the same language.

Of course, our predictions may soon be rendered obsolete. After all, it was less than a century ago that the idea of watching a YouTube video about synesthesia on an iPad while riding through the desert was unimaginable enough to be absurd. Now we have smart microwaves, turn-by-turn navigation, and twerking robots. Whatever happens, technology is sure to play a significant role in our obtaining superhuman status.

CHAPTER 10

The Transcendent Everyman

A Superhuman Mind Inside Us All

In this book we have explored the boundaries of seemingly superhuman mental ability. Our brain is capable of incredible things, but often they're only triggered or brought to the surface through accidents or rare situations. We have explored a number of cases of people who have developed extraordinary mental abilities following accidents or special circumstances or who were born with brain abnormalities. What's interesting about all of these cases is that they show us how the neurotypical brains could come to function. It shows that superhuman ability is not an ethereal property of profound individuals. It's not a gift bestowed upon the lucky few, but probably something closer to an inborn propensity that is latent in most of us.

What normally makes superhuman-level ability inaccessible to us is a combination of factors. Executive regions of our brains tend to suppress activity in more creative lower-level neural areas on the basis of cost-benefit principles, and most of us don't have ways of translating our brains' amazing abilities into something that we can interpret consciously. People with savant syndrome, on the other hand, tend to perceive literal detail such as the texture of carpet. This type of access makes them very literal perceivers. When looking at a tilted coin, they have difficulties *not* paying attention to the egocentric elliptical shape it has relative to the perceiver. Although we can all pay attention to egocentric features of objects, most people cannot retain this sort of information because it constantly changes. Savants with their prodigious memory, however, can retain literal information and use it, for example, for purposes of playing music or drawing.

But not all savants are born that way. Damage to executive areas of brains can increase activity in creative brain regions, as is seen in the case of acquired savantism. Some individuals manage to access creative brain regions via training. Such is the case for memory sports champions. With algorithms, synesthesia training, and other strategies it is possible to bring out a lot of talent right away for perfectly ordinary individuals. And we are not talking about ten thousand hours of training. We are talking about weeks or months. When we purposely internalize algorithms or synesthetic connections, we create and strengthen neural connections through the process of myelination. The initial practice makes the neurons fire, which produces a chemical known as adenosine, and adenosine binds to receptors on myelin-making cells. This makes the cells make more myelin, which speeds up neural transmission. So, eventually the axon transmits so fast that the association or movement is automatic. You no longer have to think about it for the transfer to take place. All we need is a trigger.

Once the algorithms and quasi-synesthetic associations have become internalized, there may be virtually no difference between the ordinary individual and the savant—except the ordinary individual escapes the disadvantages that may accompany a brain defect. Now *that's* smart.

As our technology improves, there are also artificial ways of teasing superhuman abilities out of people. Research done by Australian neuroscientist Allan Snyder shows that TMS, which uses electromagnetic induction to create electric currents in specific parts of the brain, can create savant-like abilities in neurotypical people.

Psychedelic drugs like psilocybin have TMS-like effects on the brain, as they melt away boundaries, allowing people to express their creative talents more freely. These drugs are currently being tested in clinical studies at hospitals around the world. Not only can a onetime exposure to these drugs bring out extraordinary creativity, it can also have positive effects on personality and mood disorders such as depression. So far, it seems that in purified form they may be the wonder drugs of the future—and possibly the very near future.

There are ways to bring out superhuman minds in people that do not require tinkering with brain functioning. With commercially available EEG and MEG (magnetoencephalography), it is already possible for us to move objects around with our minds, and it will soon be possible to text and e-mail via thought alone. It's the exciting beginning of an era in which the brains of mere mortals finally will attain the superhuman abilities formerly reserved for only the luckiest of the gods.

Larger Ramifications: Learning Disabilities

What are the larger ramifications of training our minds to be superhuman? The techniques and insights involved in this kind of training can help us deal with neurological obstacles to learning and daily function, such as learning disabilities and strokes. Although there are numerous types of learning disabilities, reading disabilities are the most common. Among children with learning disabilities, 70–80 percent have deficits in reading. Dyslexia is a common reading disability, although not the only one. It is likely the result of a deficiency in the white matter in one of the three white-matter tracts associated with phonemic awareness in reading. Interventions should therefore focus to a great extent on training with phonemes, for example, by teaching children how to recognize and identify sounds by using phonics and exercises to identify a misspelled word in a phrase and distinguish look-alikes (e.g., "innocently" and "inherently"). Training the very regions of the brain involved in the disorder can help generate myelin. Here shortcuts can speed up the learning process. A study conducted at Georgetown University Medical Center comparing dyslexic children and controls suggests that the differences in brain structure between neurotypical individuals and people with dyslexia may be the result of differences in reading experience. Poor reading experiences early on in life may trigger defects in the neural regions associated with reading. Apparently part of the malfunctioning of the brain is its neural connection to brain regions that process negative stimuli. One approach used to sever the connection between certain types of events and the negative emotions is continued exposure in a nonstressful setting. Creating positive reading experiences is thus crucial in the process of alleviating dyslexia.

Dyslexic children often have problems in math, owing to their general difficulty deciphering symbols. This is also known as dyscalculia. Transcranial magnetic stimulation, as we have seen, is one way to develop brain regions involved in math. Another way is to take advantage of their strengths. Whereas dyslexic children have problems with symbols, they tend to be good at understanding three-dimensional objects and spatial reasoning. So, the training should focus on the comprehension of symbols rather than visuospatial reasoning.

Unlike people with dyscalculia as a result of dyslexia, people with pure dyscalculia appear to have defects in a part of the parietal cortex on the top of the head involved in visuospatial reasoning, such as comparison of quantities in general, regardless of whether the quantities are physical or abstract. This then leads to deficits in number processing. TMS has indeed been used to generate temporary dyscalculia when applied over the right intraparietal sulcus, a region of the parietal cortex involved in processing physical size. The best exercises for dyscalculia then may be to train comparison of physical quantities before moving on to abstract quantities, thereby strengthening the neural regions involved in size comparison.

Stroke Recovery

Stroke, the fourth-highest killer of Americans and the leading cause of adult disability, is a debilitating injury that very often deprives individuals of mobility as well as the ability to speak. You lose around two million brain cells a minute when you are having an ischemic stroke, a blood clot in the fleshy insides of your skull. Each hour that the stroke continues is equivalent to 3.6 years of natural brain aging. Given that the average untreated ischemic stroke lasts for ten hours, fast intervention is key to recovery. But even with fast intervention, stroke recovery often requires serious reorganization or retraining of the brain to regain mobility and language. Fortunately, stroke survivors can learn a lot from how cognitive skills can be implemented in neurotypical brains. One insight into brain plasticity that is already widely used in stroke recovery is that mere *thought* can lead to a rewiring of nearly all brain regions. In a study conducted by Pascual-Leone, participants learned and practiced a little five-finger piano exercise. They practiced two hours a day for five days. Using TMS, which can be used to interfere with brain activity in a localized region as we saw in chapter 9, but can also be used to measure brain activity, the researchers verified that parts of the motor cortex were active during the piano exercise. This, of course, was expected, as we know that the motor cortex is involved in movement. But in a second follow-up study, new participants were asked to merely *think* about playing the piano instead of actually playing, holding their hand still while imagining their fingers moving up and down. Researchers found that the region of the motor cortex that controls the piano-playing fingers also expanded in the brains of volunteers who imagined playing the music—just as it had in those who actually played it. Thus, even practicing a physical activity in your head can help rewire the brain to be able to complete the activity. So, while the brain of a stroke patient starts to heal and the limbs are unmovable, patients can still train the brain regions affected by the condition by imagining moving their limbs.

Strokes in the left temporal lobe can knock out the patient's language abilities, in either the short term or long term. In the case of a massive stroke in the left hemisphere, the brain is likely to move the language center from the left to the right hemisphere during recovery. There are good reasons for that. The damaged region in the left hemisphere might not work at all while it is healing. So, language training makes use of different brain regions to perform the task. The problem is that the right hemisphere tends not to pick up language skills as well as the left hemisphere. Therefore, the stroke survivor typically ends up with impoverished language functions. What we know about the brains of acquired savants can help with that problem. Acquired savants typically have damages to the dominant hemisphere, which then allows skills in the other hemisphere to develop. We have already seen that TMS can be used to simulate a lesion in one hemisphere, allowing the other hemisphere to bloom. If we use TMS to inhibit the right hemisphere during stroke recovery, we can force the left hemisphere to pick up the language functions it lost. Clinical trials are currently being conducted, testing this approach, and the initial results look promising. Language recovery in the left hemisphere tends to restore more of the original functions than language development in the right hemisphere.

But you don't have to wait for TMS to be FDA approved for stroke recovery to take advantage of this insight. As we have seen in the case of artistic skills, we can teach the brain to switch from a left-hemisphere mode to a right-hemisphere mode. There are also ways to teach the brain to switch from a right-hemisphere mode to a left-hemisphere mode. Whereas the right hemisphere processes literal detail, the left hemisphere is dominant in abstracting away from literal detail and computing a bigger picture. So, abstraction and interpretation exercises can help switch to a left-hemisphere mode of thinking. But to minimize the chance that the brain switches from left to right, abstraction and interpretation exercises must avoid the use of language. Find-the-missing-number exercises are particularly effective because the brain processes math in the left hemisphere. Here are a few examples:

32, 25, ?, 11
3, 3, 6, 12, ?
2, 6, 14, ?, 62

The solutions to these are: 18 (subtract 7), 24 (add up all previous numbers, $3 + 3 = 6$; $3 + 3 + 6 = 12$; $3 + 3 + 6 + 12 = 24$), and 30 (multiply by 2 and add 2).

Spot-the-difference exercises also help the brain switch its mode from the right to the left hemisphere. This type of exercise requires comparing two pictures to one another in order to identify the differences.

Another type of exercise that involves the left hemisphere is the formation of novel visual imagery. For example, try to visualize a novel object that is composed of your iPhone and your reading glasses, or a novel creature that is composed of your dog and a chicken.

Ideally these types of exercises should eventually be supplemented by abstraction exercises that use language, such as the following:

What word do you think is related to the following sets of three words?

> Hunt, air, stand
> Answer: *head . . . headhunt, airhead,* and *headstand.*
> Step, rest, ball
> Answer: *foot . . . footstep, footrest,* and *football.*
> Inter, down, pre
> Answer: *face . . . interface, facedown, preface*

The mix of language abstraction exercises and abstraction exercises that do not rely on language may help restore the language center in the left hemisphere, thus avoiding the defects that remain when the right brain takes over.

Of course, there may be costs associated with keeping the left hemisphere dominant: Some stroke survivors become prodigious savants as a result of a stroke. Such was the case for Ken Walters, an engineer. When he woke up in the hospital after his stroke, he immediately began to doodle. Without even being aware of what he was doing, Ken had drawn something truly amazing. Although he had sincerely disliked art in school, he now could not stop thinking about it and practicing. Ken's enhanced right-brain skills were a result of brain damage and may not have developed if a left-hemisphere mode had been imposed on his brain. And Ken is in no way unique. Former troublemaker Tommy McHugh started writing poetry, painting, and sculpting after he had brain aneurysms in his executive frontal lobe regions. He soon became famous for his alien-like doodles and has had his art exhibited in many galleries in the UK. And Jon Sarkin, a chiropractor, became obsessed with painting after he had part of his brain removed after a stroke, and like the others he had never had any talent or interest in art prior to the incident. His paintings now regularly sell for thousands of dollars.

Using More of Your Brainpower

But you can not only use the insights from people with acquired superhuman mental abilities to learn how to recuperate from brain damage, you can also make the best use of your brain even when you are afflicted with any apparent disability. While practice is needed to acquire a new skill, the right kind of practice can make the reacquisition of the skill feasible by shaving off hours of useless brain exercise. You can help the brain form new connections by using the right sort of approach and a suitable algorithm. Many of the skills acquired when practicing a narrow skill are widely applicable. An example of this is learning to memorize pi. When digits and short strings of digits take on meaning, they will be easier to remember when they occur elsewhere as well. If 177 is a meaningful part of a narrative, then upon hearing that your flight is taking off from Gate C177, you no longer need to stop every five minutes to recheck the gate information on your boarding pass. The information will stick right away. More generally, by practicing memory in one domain, your ability to store and retrieve information will improve in general, making you a better learner.

Another example of how a skill in a narrow area can be more widely applicable is that of perfect pitch. Because tone languages, such as Mandarin and Vietnamese, require hitting the right pitch, it can be difficult for people who didn't grow up speaking a tone language to acquire a tone language later in life. They lack the perfect pitch of many of the people who grew up speaking a tone language. Acquiring perfect pitch, or even almost perfect pitch, means eliminating one of the major obstacles standing in the way of learning to speak a tone language.

By practicing superhuman abilities, you develop more and stronger brain connections throughout the brain. It is sometimes said that we use only 10 percent of our brain. This claim is false. Of course it is. The point of the pruning processes that take place during childhood is to get rid of superfluous brain connections and strengthen the ones we use. So, we use all three pounds of tissue in our heads. If a part of the brain is not used, it will deteriorate and disappear.

What is true is a claim in the close vicinity of the myth. The number of brain connections we use on a daily basis can vary greatly, depending on the activities we engage in. The brain's capacity, or power, is its ability to expand its neurons, generate new neural connections, and strengthen existing ones. Only the future will be able to tell how powerful our brains are. But it is unlikely that we already utilize as much as 10 percent of our brain's power. We likely utilize considerably less. So, while it is false that a hope for the future is to learn how to use more of our brain, it is true that many of us hope to discover new and better ways to use more of our brain's *power.*

The Abacus

The abacus allows one to visually represent the way numbers are cal-culated. Consider what happens when you count to twenty in your head, starting at 0:

0, 1, 2, 3, 4, 5, 6, 7, 8, 9, 10, 11, 12, 13, 14, 15, 16, 17, 18, 19, 20

We've all become so accustomed to counting that we often don't think about how counting actually works. If you include some leading zeros, you'll see what is actually happening:

00, 01, 02, 03, 04, 05, 06, 07, 08, 09, 10, 11, 12, 13, 14, 15, 16, 17, 18, 19, 20

The first column from the right (the ones column) is updated every time you advance one number, but the second column from the right (the tens column) only advances every time your count advances ten places. Each column of a number corresponds to a value representing ten times the value of the column to its right. Counting to higher numbers like 1000 follows the same process:

0000, 0001, 0002, . . . , 0009, 0010, 0011, . . . ,
0099, 0100, 0101, . . . , 0999, 1000

The abacus uses a similar method of keeping track of numbers. In the Japanese version—also known as a soroban—there are two rows of beads distributed in an odd number of columns—at least nine. Each column in the top row contains one bead worth 5, and each column in the bottom row contains four beads, each worth 1.

Each whole column represents a different place value of a number. The rightmost column represents the ones. It's able to store values from 1 to 9. The one next to that represents the tens place ranging from 10 to 90, and so on. So if you want to represent the number 4, move four beads in the lower ones column. If you want to represent 14, add one bead in the lower tens column. To represent 54, reset the tens column, then move the bead in the top tens column.

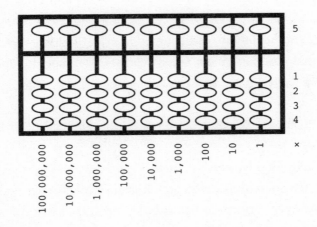

Calculations are performed by moving the bead around in a particular pattern so that the new arrangement of the beads is the result of the calculation. Start with all the bottom beads down and the top beads up. Moving a bead up in the lower row is adding and moving a bead down subtracting. It's the other way around on the top row. If you want to add 123,456 and 111,111, you will need to use the six farthest right columns. You move one bead from the lower row in column 6 (counting from right to left), two beads up from the lower row in column 5, three beads from the lower row up in column 4, all the lower beads up in column 3, one bead up from the *top* row in column 2, and one bead up from the lower and from the upper row in column 1. Now we add 111,111. Start by moving one lower bead up in columns 6–4. When you get to column 3 you cannot move a lower bead up, since all the lower beads are already up. Since 5 minus 4 is 1, you move the bead down in the top column and all four bottom beads down. In columns 2 and 1 you move one bead up. In column 1 that will be bead number 3 counting from the bottom up.

Now you can read off the result from the abacus. In column 6 you will find that two beads are up, three are up in column 5, and four are up in column 4. One upper bead is down in column 3. In column 2 one top bead is down and one lower bead is up, and in column 1 one upper bead is down and two lower beads are up. So, the result is 234,567. Of course, this particular addition would have been easy to calculate in your head without using an abacus. But the same method can be used to perform much more complicated additions, subtractions, and even multiplications.

The Doomsday Method

For this method you need to know that leap years take place every four years. For example, 2012 was a leap year. The two most recent leap years are highlighted in bold below. With this method you internalize a list of the weekdays of the first day of each year in the range of years for which you want to perform calendar calculations.

2005	2006	2007	**2008**	2009	2010	2011	**2012**	2013	2014
Sat	Sun	Mon	Tue	Thu	Fri	Sat	Sun	Tue	Wed

The first day of each year is fairly easy to remember, as each year moves forward one day, except when the previous year is a leap year. In that case it moves forward two days. So, since 2004 was a leap year and the first day of 2004 was a Thursday, the first day of 2005 was a Saturday rather than a Friday.

With this method you also need to internalize the number of days in each month. This is exceptionally easy. Start with your left hand's little finger's knuckle. That is a mountain, indicating 31 days. Then move to the valley between the little finger and the ring finger on your left hand. That is February. February is special. It has 28 days in a non–leap year and 29 days in a leap year. Then move to the knuckle on the ring finger. That is March, a tall mountain—that shows that March has 31 days. You move through all the fingers. Mountains indicate 31 days and valleys 30 days, except February. When you run out of fingers on one hand, you start again on the same hand or alternatively the other hand. As both July and August are knuckles (or mountains rather than valleys), July and August both have 31 days.

The general algorithm is to add the days in the months preceding the target month and the days in the target month, subtract 1, divide by 7, and then count forward from the first day of that year on the basis of the remainder. Let's consider some examples:

January 15, 2005. January 15 is 14 days (or two weeks) ahead of the first day of 2005. According to the above table, the first day of 2005 was a Saturday. So, January 15, 2005, was a Saturday.

February 27, 2010. The first day of 2010 was a Friday. The number of days in January is 31 and the number of days in the target month was 27. Adding the two equals 58. Subtract 1. That's 57. Divide by 7. That's 8 weeks and a remainder of 1. As the first day of 2010 was a Friday, February 27 was a Saturday.

March 5, 2012. 2012 was a leap year. 31 (January) plus 29 (February) plus 5 (March) equals 65. Subtract 1. That's 64. Divide by 7. That's 9 plus 1. As the first day in 2012 was a Sunday, March 5 was a Monday.

May 30, 2014. 31 (January) plus 28 (February) plus 31 (March) equals 80. This divided by 7 equals 11 weeks and 3 days. As the first day of 2014 was a Wednesday, May 30 was a Saturday.

We can make this even simpler by internalizing the different sums of months. This is summarized in the following table:

JAN	FEB	MAR	APR	MAY	JUN	JUL	AUG	SEP	OCT	NOV	DEC
0	31	59	90	120	151	181	212	243	273	304	334

Now it is super-easy to do calendar calculation. The general algorithm is to add the days in the target date to the number in the table for the month in question. Subtract 1. Divide by 7. Add the remainder to the first day of the year in question. Let us illustrate:

December 7, 2011. 334 + 7 equals 341. We subtract 1 and get 340. This divided by 7 yields 48 weeks and 4 days. As the first day of 2011 was a Saturday, December 7 was a Wednesday.

One problem with this method is that it requires fast division by 7. We can, however, avoid that. The table below shows the numbers from the previous table divided by 7, with their remainders. In the third row the remainder is repeated.

Jan	Feb	Mar	Apr	May	Jun	Jul	Aug	Sep	Oct	Nov	Dec
0	4+3	8+3	12+6	17+1	21+4	25+6	30+2	34+5	39+0	43+3	47+5
0	3	3	6	1	4	6	2	5	0	3	5

The only information you need to remember is the number in the last row of the table above, as well as the information from the first table (p. 236). The general algorithm proceeds as follows: Locate the remainder for the target month and add the number of days in the target month. Subtract 1, unless the year is a leap year. If the year is a leap year, don't subtract anything. Use the result to count forward from the first day of the year. Let's illustrate:

November 3, 2009. Add the remainder for November, which is 3 (third table), to the target date 3 (November 3). That is 6. Subtracting 1 yields 5. The first day of the year in 2009 was a Thursday (first table). Counting 5 days forward yields a Tuesday. So, November 3, 2009, was a Tuesday.

October 15, 2008. The remainder for October is 0 (second table). Add that to 15 (October 15). This yields 15. As 2008 was a leap year, we don't subtract anything. The first day of the year 2008 was a Tuesday (first table). Count 15 days forward (or 1, because $(2 \times 7) + 1$ equals 15). That yields a Wednesday. So, October 15, 2008, was a Wednesday.

September 8, 2005. Add the remainder for September, which is 5 (second table), to the target date 8 (September 8). That yields 13. We subtract 1. That yields 12. The first day of the year 2005 was a Saturday (first table). Count 12 days forward. That yields a Thursday. So, September 8, 2005, was a Thursday.

July 17, 2008. Add the remainder for July, which is 6 (second table), to the target date 17 (July 17). That yields 23. As 2008 was a leap year, we don't subtract anything. The first day of the year 2008 was a Tuesday (first table). Count 23 days forward (or 2, because $(3 \times 7) + 2$ yields 23). That yields a Thursday. So, July 17, 2008, was a Thursday.

If you want to do calendar calculation that goes far beyond a span of ten years, things again become a bit more complicated, but with practice the calculations can be carried out in a few seconds. Let us focus on the time span 1800 to 2100. We begin by memorizing a number between 0 and 6 assigned to the days of the week, starting with Sunday. So, Sunday is 0 and Saturday is 6. We then memorize the following *anchor days* for the four centuries we are focusing on:

1800–1899: Friday
1900–1999: Wednesday
2000–2099: Tuesday
2100–2199: Sunday

We furthermore memorize the following month-and-number associations—the so-called doomsdays:

JAN	FEB	MAR	APR	MAY	JUN	JUL	AUG	SEP	OCT	NOV	DEC
3*	28*	7	4	9	6	11	8	5	10	7	12

Years divisible by 4 are leap years, *unless* the year is divisible by 100. Then it is not a leap year, unless the year is also divisible by 400. Then it *is* a leap year. So, the year 2004 was a leap year, because it is divisible by 4 but not 100. The year 1500 is divisible by 4 and also by 100, but not by 400. So, 1500 was *not* a leap year. The year 1600 is divisible by 4 and 100 but also by 400. So, 1600 was a leap year. In leap years the number for January is 4 and the number for February is 29. The asterisks above signify that the number for January and February changes, depending on whether the year in question is a leap year.

Most of these dates are super-easy to remember. For example, 4/4, 5/9 (reverse work hours), 6/6, 7/11 (convenience store), 8/8, 9/5 (work hours), 10/10, 11/7 (reverse convenience store), 12/12. The only months for which the doomsdays deviate from this pattern are January (3*) and February (28*).

The method now is to calculate how many multiples of 12 fit in the last two digits of the target year, as wells as the remainder. Then calculate how many multiples of 4 fit into the remainder. Locate the anchor day for the century and its number (e.g., Friday is 5). Add up the four numbers. Subtract whole multiples of 7 and identify the remainder. Find the day of the week for the remainder (e.g., 0 is Sunday). That day is also the day of the doomsday. Identify the doomsday target month in the doomsday table. Count forward to the target date. Let's look at some examples:

April 12, 1808: 12 fits 0 whole times into the last two digits (08) of the target year. So, 8 is the remainder. 4 (April) fits 2 whole times into the remainder of 8. The anchor day for the century 1800 is Friday, which is day 5 in the week. Add the four numbers: 0 + 8 + 2 + 5 = 15. 7 fits 2 whole times into 15 with 1 as a remainder. 1 is a Monday. So, the doomsday for 1808 is a Monday. Identify the doomsday for the target month—April—in the doomsday table above. That's April 4. So, April 4, 1808, was a Monday. Count forward from April 4, 1808 (Monday), to April 12, 1808. Since April 4, 1808, was a Monday, April 12, 1808 (8 days later), was a Tuesday. So, April 12, 1808, was a Tuesday.

September 21, 1955. 12 times 4 is 48. So, 12 fits 4 whole times into the last two digits (55) of the target year. The remainder (55–48) is 7. 4 fits 1 whole time into 7. The anchor day for the century 1900 is Wednesday, which is day 3 in the week. Add the four numbers: 4 + 7 + 1 + 3 = 15. 7 fits 2 whole times into 15 with a remainder of 1. Day 1 of the week is Monday. So, the doomsday for 1955 is Monday. The doomsday for the target month September is September 5 (work hours, 9/5). So, September 5, 1955, was a Monday. Count forward from September 5, 1955 (Monday), to September 21, 1955. That's 2 weeks and 2 days (21 minus 5 equals 16). Counting forward 2 weeks and 2 days from Monday yields a Wednesday. So, September 21, 1955, was a Wednesday.

August 28, 2015: 12 times 1 equals 12. So, 12 fits 1 whole time into the last two digits (15) of the target year. The remainder (15 − 12) is 3. 4 fits 0 times into 3. The anchor day for the century 2000 is Tuesday, which is day 2 in the week. Adding the numbers 1, 3, 0, and 2 yields 6. 7 fits 0 whole times into 6, with a remainder of 6. Day 6 of the week is Saturday. So, August 8, 2015 (a doomsday), was a Saturday. Counting forward from August 8, 2015 (Saturday), to August 28, 2015, yields 2 weeks and 6 days (28 minus 8 yields 20). Counting forward 2 weeks and 6 days yields a Friday. So, August 28, 2015, was a Friday.

The Gregorian calendar repeats exactly every four hundred years. So, what we just went through can be extended to all centuries, as long as we use the Gregorian calendar, which was not introduced until 1582. So, there will be discrepancies between dates in the Gregorian calendar and the inaccurate calendar used before. (The Julian calendar introduced an error of one day every 128 years. The Gregorian calendar introduces an error of one day every 3,236 years.) Let us look at some examples of Gregorian calendar calculation for other centuries:

July 4, 1776: 12 times 6 equals 72. So, 12 fits 6 whole times into the last two digits (76) of the target year. The remainder (76 − 72) is 4. 4 fits 1 whole time into 4. The anchor day for the century 1700 is Sunday, which is day 0 in the week. Adding the numbers 6, 4, 1, and 0 yields 11. 7 fits 1 whole time into 11, with a remainder of 4. Day 4 of the week is Thursday. So, July 11, 2015 (a doomsday), was a Thursday. July 11, 1776 (Thursday), is 7 days later than July 4, 1776. So, July 4, 1776, was a Thursday.

March 3, 1699: 12 times 8 equals 96. So, 12 fits 8 whole times into the last two digits (99) of the target year. The remainder (99 − 96) is 3. 4 fits 0 times into 3. The anchor day for the century 1600 is Tuesday, which is day 2 in the week. Adding the numbers 8, 3, 0, and 2 yields 13. 7 fits 1 whole time into 13, with a remainder of 6. Day 6 of the week is Saturday. So, March 7, 1699 (a doomsday), was a Saturday. March 7, 1699, is 4 days later than March 3. Counting 4 days backward from Saturday gives us a Tuesday. So, March 3, 1699, was a Tuesday.

NOTES

Chapter 1: The Hidden Abilities in All of Us

10 **mental feats related to their synesthetic abilities:** Rothen N, Meier B, Ward J. Enhanced memory ability: Insights from synaesthesia. *Neuroscience & Biobehavioral Reviews*. 2012; 36(8): 1952–63.

14 **whether it is, say, an E or a D:** Deutsch D. The puzzle of absolute pitch. *Current Directions in Psychological Science*. 2002; 11(6): 200–204.

15 **Julia Simner explained to the BBC:** http://news.bbc.co.uk/2/hi/8248589.stm

17 **involved in his newfound skills:** Discovery Channel, 2003.

19 **rise to cognitive and verbal disinhibition:** Lythgoe MFX, Pollak TA, Kalmus M, de Haan M, Chong WK. Obsessive, prolific artistic output following subarachnoid hemorrhage. *Neurology*. 2005; *64*: 397–98.

Chapter 2: Brilliant Impact

29 **"because the fluid appeared to be frozen, like a glacier":** Zihl J, von Cramon D, Mai N. Selective disturbance of movement vision after bilateral brain damage. *Brain*. 1983; *106*: 313–40.

32 **Jason later proved that his drawing represented pi:** Brogaard B. A case of acquired savant syndrome and synesthesia following a brutal

assault. CAS Grant Report, University of Missouri, St. Louis. 2011. Padgett J, Seaberg M. *Struck by Genius: How a Brain Injury Made Me a Mathematical Marvel.* New York: Houghton Mifflin Harcourt; 2014.

33 **impossible mathematical, linguistic, or artistic tasks:** Treffert DA. The savant syndrome: An extraordinary condition. A synopsis: Past, present, future. *Philosophical Transactions of the Royal Society.* 2009; *B* 27(364): 1351–57.

34 **dominates the average person's brain activity:** Snyder AW, Mulcahy E, Taylor JL, et al. Savant-like skills exposed in normal people by suppressing the left fronto-temporal lobe. *Journal of Integrative Neuroscience.* 2003; 2(2): 149–58. Young RL, Ridding MC, Morrell TL. Switching skills by turning off part of the brain. *Neurocase.* 2004; *10*(3): 215–22.

36 **Garfford Broussard:** Garfford Broussard was born in 1921, a time that, according to Gladwell, should have made it difficult to succeed. However, he still offers a powerful illustration of Gladwell's overall idea, for love and luck no doubt played a large role in Broussard's career.

38 **extensive research of how the brain processes mathematics:** Dehaene S. *The Number Sense.* Oxford: Oxford University Press; 1999. Dehaene S. Précis of The Number Sense. *Mind & Language.* 2001; *16*(1): 16–36. Dehaene S, Dehaene-Lambertz G, Cohen L. Abstract representations of numbers in the animal and human brain. *Trends in Neurosciences.* 1998; *21*(8): 355–61. Dehaene S, Molko N, Cohen L, Wilson A. Arithmetic and the brain. *Current Opinion in Neurobiology.* 2004; *14*: 218–24.

39 **this theory of how the brain processes numbers:** Dehaene S. *The Number Sense.* Oxford: Oxford University Press; 1999. Dehaene S. Précis of The Number Sense. *Mind & Language.* 2001; *16*(1): 16–36.

39 **based on what he calls the Doomsday Rule:** Conway JH. Tomorrow is the day after doomsday. *Eureka.* 1973; *36*: 28–31. Berlekamp E, Guy R, Conway JH. *Winning Ways for Your Mathematical Plays, Vol. 2.* London: American Press; 1982.

40 **a few dates every time he logs on to his computer:** Alpert M. Not Just Fun and Games. *Scientific American.* April 1999.

45 **Research Unit and Magnetic Imaging Centre at Aalto University in Finland:** Brogaard B, Vanni S, Silvanto J. Seeing mathematics: perceptual experience and brain activity in acquired synesthesia. *Neurocase*. 2013; *19*(6): 566–75.

45 **arbitrariness among particular synesthetic experiences:** Eagleman DM, Kagan AD, Nelson SS, Sagaram D, Sarma AK. A standardized test battery for the study of synesthesia. *Journal of Neuroscience Methods*. 2007; *159*(1): 139–45.

46 **mostly on that side of the brain:** Krueger F, Spampinato MV, Pardini M, et al. Integral calculus problem solving: an fMRI investigation. *NeuroReport*. 2008; *19*: 1095–99.

46 **mostly a left-hemisphere activity:** Yomogida Y, Sugiura M, Watanabe J, et al. Mental visual synthesis is originated in the fronto-temporal network of the left hemisphere. *Cerebral Cortex*. 2004; *14*(12): 1376–83.

47 **imagery that doesn't exist in the real world:** Yomogida Y, Sugiura M, Watanabe J, et al. Mental visual synthesis is originated in the fronto-temporal network of the left hemisphere. *Cerebral Cortex*. 2004; *14*(12): 1376–83.

47 **increased activation in certain parts of the visual cortex:** Aleman AA, Rutten GM, Sitskoorn MM, Dautzenberg G, Ramsey NF. Activation of striate cortex in the absence of visual stimulation: an fMRI study of synesthesia. *NeuroReport*. 2001; *12*: 2827–30. Nunn JA, Gregory LJ, Brammer M, et al. Functional magnetic resonance imaging of synesthesia: activation of V4/V8 by spoken words. *Nature Neuroscience*. 2002; *5*: 371–75.

52 **acumen may manifest without training:** Koelsch S, Gunter T, Friederici AD, Schroèger E. Brain indices of music processing: "nonmusicians" are musical. *Journal of Cognitive Neuroscience*. 2000; *12*(3): 520–41.

52 **motor function, sensation, and language processing:** Gaser C, Schlaug G. Brain structures differ between musicians and nonmusicians. *The Journal of Neuroscience*. 2003; *23*(27): 9240–45.

52 **facilitate the demands of musical performance:** Croom AM. Music, neuroscience, and the psychology of well-being: a précis. *Frontiers in Theoretical and Philosophical Psychology*. 2012; *2*(393): 1–15.

53 **efficiently, using minimal resources:** Krings T, Töpper R, Foltys H, et al. Cortical activation patterns during complex motor tasks in piano players and control subjects. A functional magnetic resonance imaging study. *Neuroscience Letters.* 2000; *278*(3): 189–93.

54 **a woman developed synesthesia after a stroke:** Ro T, Ellmore TM, Beauchamp MS. A neural link between feeling and hearing. *Cerebral Cortex.* 2012; 23: 1724–1730. doi: 10.1093/cercor/bhs166

58 **outgrowth of abnormal neural connections**: Giza CC, Prins ML. Is being plastic fantastic? mechanisms of altered plasticity after developmental traumatic brain Injury. *Developmental Neuroscience.* 2006; *28*: 364–79. Konrad C, Geburek AJ, Rist F, et al. Long-term cognitive and emotional consequences of mild traumatic brain injury. *Psychological Medicine.* 2011; *41*(6): 1197–211.

Chapter 3: A Flexible Mind

59 **written or been able to connect to the world:** http://carlysvoice.com/home/aboutcarly/
https://www.youtube.com/watch?v=34xoYwLNpvw

60 **widely believed in the past:** Markram H, Markram K. Interview: Henry and Kamila Markram about The Intense World Theory for Autism. Holman JS (Interviewer), 2013. WrongPlanet.net. http://www.wrongplanet.net/article419.html. Markram K, Markram H. The Intense World Theory—A Unifying Theory of the Neurobiology of Autism. *Frontiers in Human Neuroscience.* 2010; *4.*

61 **moderate to severe autism as a young child:** http://www.cnn.com/2014/04/29/health/irpt-autism-communicating/

62 **he was in kindergarten—and autistic:** http://www.scientificamerican.com/article/the-hidden-potential-of-autistic-kids/

62 **abnormality in the serotonin system:** Brogaard B. Serotonergic hyperactivity as a potential factor in developmental, acquired and drug-induced synesthesia. *Frontiers in Human Neuroscience.* October 2013: *21.* doi: 10.3389/fnhum.2013.00657

63 **serotonin-increasing antidepressant drugs during pregnancy:** http://canadajournal.net/health/anti-depressants-pregnancy-linked -autism-study-6323-2014/

65 **three times as likely to be diagnosed with autism:** http://canada journal.net/health/anti-depressants-pregnancy-linked-autism-study -6323-2014/

67 **at San Diego's East County Performing Arts Center:** https://www .youtube.com/watch?v=2_zWuPeSDs0&feature=youtu.be

68 **genes on chromosome 7:** http://www.kentucky.com/2011/05/11/1736354 /for-woman-with-williams-syndrome.html

69 **found the best voice teachers they could:** http://articles.latimes.com /1990-12-23/news/vw-9838_1_gloria-lenhoff

72 **observing neurons in a microscope:** Munz M, Gobert D, Schohl A, et al. Rapid Hebbian axonal remodeling mediated by visual stimulation. *Science.* 2014; *344* (6186): 904. doi:10.1126/science.1251593

80 **largeness relative to the objects that surround it:** Massaro DW, Anderson NH. Judgmental model of the Ebbinghaus illusion. *Journal of Experimental Psychology.* 1971; 89(1), 147–51.

81 **automatically calculating unconscious brain:** Goodale MA, Milner AD. Separate visual pathways for perception and action. *Trends in Neurosciences.* 1992; *15*(1): 20–25. Milner AD, Goodale MA. *The Visual Brain in Action.* New York: Oxford University Press; 1995.

Chapter 4: What's Your Number?

86 **corpus callosum suffered from split-brain syndrome:** Gazzaniga M, LeDoux JE. *The Integrated Mind.* London: Plenum Press; 1978.

89 **argued that memory is limited to nine items:** Miller GA. The magical number seven, plus or minus two: Some limits on our capacity for processing information. *Psychological Review.* 1956; *63*: 81–97.

89 **and it fades quickly:** Baddeley AD. The episodic buffer: a new component of working memory? *Trends in Cognitive Sciences.* 2000; *4*(11): 417–23.

90 **high-functioning brain and a learning disability:** Jacobson LA, Ryan M, Martin RB, et al. Working memory influences processing speed and reading fluency in ADHD. *Child Neuropsychology.* 2011; *17*(3): 209–24.

91 **memories of specific events:** http://www.npr.org/blogs/health/2014/04/08/299189442/the-forgotten-childhood-why-early-memories-fade

95 **Synesthesia can aid memory:** Rothen N, Meier B, Ward J. Enhanced memory ability: Insights from synaesthesia. *Neuroscience & Biobehavioral Reviews.* 2012; *36*: 1952–63.

95 **what they call quasi-synesthesia can be learned:**http://www.plosone.org/article/info%3Adoi%2F10.1371%2Fjournal.pone.0039799

96 **neuroscientists Vilayanur S. Ramachandran and Edward Hubbard:** Ramachandran VS, Hubbard EM. Psychophysical investigations into the neural basis of synaesthesia. *Proceedings of the Royal Society.* 2001a; *268*(1470): 979–983. doi:10.1098/rspb.2000.1576

98 **contributing to the enhancement of memory:** Chwilla DJ, Brunia CH. Effects of emotion on event-related potentials in an arithmetic task. *Journal of Psychophysiology.* 1992; *6*(4): 321–32.

103 **names are the most difficult to get right:** Griffin ZM. Retrieving Personal Names, Referring Expressions, and Terms of Address. In: Ross BH, ed. *The Psychology of Learning and Motivation, Vol. 53.* Burlington, IN: Academic Press; 2010: 345–87.

103 **the tip-of-the-tongue phenomenon (TOT):** Brown R, McNeill D. The "tip-of-the-tongue" phenomenon. *Journal of Verbal Learning and Verbal Behavior.* 1966; *5*: 325–37.

103 **but hadn't seen in a while:** Burke DM, MacKay DG, Worthley JS, Wade E. On the tip of the tongue: What causes word finding failures in young and older adults? *Journal of Memory and Language.* 1991; *30*(5): 542–79.

104 **with the individuals they are naming:** Of course, like everything, this has been the subject of debate. See Allerton DJ. The linguistic and sociolinguistic status of proper names: What are they, and who do they belong to? *Journal of Pragmatics,* 1987; *11*(1), 61–92. Levi-Strauss C. *The Savage*

Mind. Chicago: University of Chicago Press; 1966. Miller GA, Johnson-Laird PN. *Language and Perception.* Cambridge, MA: Harvard University Press; 1976.

104 **remember a name if it's *not* so unique:** James LE, Fogler KA. Meeting Mr. Davis vs Mr. Davin: Effects of name frequency on learning proper names in young and older adults. *Memory.* 2007; *15*(4): 366–74.

104 **the person standing in front of you:** Brennen T. The difficulty with recalling people's names: The plausible phonology hypothesis. In: Cohen G, Burke DM, eds. *Memory for Proper Names.* Hillsdale, NJ: Lawrence Erlbaum Associates; 1993: 409–31.

104 **remembering her phone number and address:** Harris DM, Kay J. I recognize your face but I can't remember your name: Is it because names are unique? *British Journal of Psychology.* 1995; *86*(3): 345–58. Saetti MC, Marangolo P, De Renzi E, Rinaldi MC, Lattanzi E. The nature of the disorder underlying the inability to retrieve proper names. *Cortex.* 1999; *35*(5): 675–85.

104 **helps you remember the name:** Brooks JO III, Friedman L, Gibson JM, Yesavage JA. Spontaneous mnemonic strategies used by older and younger adults to remember proper names. In: Cohen G, Burke DM, eds. *Memory for Proper Names.* Hillsdale, NJ: Lawrence Erlbaum Associates; 1993: 393–407. Milders M, Deelman B, Berg I. Rehabilitation of memory for people's names. *Memory.* 1998; *6*(1): 21–36. Morris PE, Fritz CO, Jackson L, Nichol E, Roberts E. Strategies for learning proper names: Expanding retrieval practice, meaning and imagery. *Applied Cognitive Psychology.* 2005; *19*(6): 779–98.

106 **mental models of a series of spoken directions:** Lee PU, Tversky B. Interplay between visual and spatial: The effect of landmark descriptions on comprehension of route/survey spatial descriptions. *Spatial Cognition and Computation.* 2005; *5*(2&3): 163–85. Mani K, Johnson-Laird PN. The mental representation of spatial descriptions. *Memory & Cognition.* 1982; *10*(2): 181–87. Taylor HA, Tversky B. Spatial mental models derived from survey and route descriptions. *Journal of Memory and Language.* 1992; (31): 261–82.

106 **keep them as vivid as possible:** Tom AC, Tversky B. Remembering routes: Streets and landmarks. *Applied Cognitive Psychology.* 2012; 26(2): 182–93.

108 **two to three years at any college:** Woodsmall M, Woodsmall W. *The Future of Learning: The Michel Thomas Method.* Next Step Press; 2008: 142.

111 **stronger emotional bonds than those who don't:** Vallotton CD, Ayoub CC. Symbols build communication and thought: The role of gestures and words in the development of engagement skills and social-emotional concepts during toddlerhood. *Social Development.* 2010; 19(3): 601–26.

111 **verbal communication skills during later years:** Goodwyn SW, Acredolo LP, Brown CA. Impact of symbolic gesturing on early language development. *Journal of Nonverbal Behavior.* 2000; 24: 81–103.

112 **higher IQs, to the tune of twelve points on average:** Acredolo LP, Goodwyn SW. The longterm impact of symbolic gesturing during infancy on IQ at age 8. Paper presented at *International Conference on Infant Studies.* Brighton, UK; 2000.

Chapter 5: Smart Cookies

117 **the abacus and the left side for the numbers:** You can see a video of the children demonstrating this here: http://youtube/_hDwP5F32wc

Chapter 6: Draw Like a Child

127 **as a result of frontotemporal dementia:** Miller BL, Ponton M, Benson DF, Cummings JL, Mena I. Enhanced artistic creativity with temporal lobe degeneration. *Lancet.* 1996; 348(9043): 1744–45.

133 **wants to attach meaning to everything it sees:** A number of useful ones can be found in: Edwards B. *Drawing on the Right Side of the Brain.* New York: Jeremy P. Tarcher/Putnam; 1979.

134 **Betty Edwards calls the "picture plane":** Edwards B. *The New Drawing on the Right Side of the Brain.* New York: Jeremy P. Tarcher/Putnam; 1999.

Chapter 7: Not a Worry in the World

140 **We call it sleepwalking:** Sleepwalking is also known as somnambulism or noctambulism.   .

140 **have sex with strangers:** Nowak R. Sleepwalking woman had sex with strangers. *New Scientist.* October 15, 2004. http://www.newscientist.com/article/dn6540-sleepwalking-woman-had-sex-with-strangers.html#.VCx6PRZlHkc

140 **spaghetti Bolognese, omelets, and fish and chips during his sleep:** Scott K. Sleepwalking chef's recipe for disaster. *The Guardian.* March 30, 2006. http://www.theguardian.com/uk/2006/mar/31/kirstyscott.uknews2

140 **wrote the solutions to various problems in her sleep:** Woman inventor. *Western Age* (Dubbo, New South Wales). August 28, 1915: 4.

141 **because we get sleepy," he says:** Max DT. The secrets of sleep. *National Geographic Magazine.* 2010. http://ngm.nationalgeographic.com/2010/05/sleep/max-text

141 **find animals that don't need to sleep at all:** Chiara C, Tononi G. Is sleep essential? *PLOS Biology.* 2008; 6(8): e216.

142 **net loss of energy for each sleep period:** Daan S, Barnes BM, Strijkstra AM. Warming up for sleep? Ground squirrels sleep during arousals from hibernation. *Neuroscience Letters.* 1991; 128(2): 265–68.

142 **faster in the sleeping state than in the waking state:** Xie L, Kang H, Xu Q, et al. Sleep drives metabolite clearance from the adult brain. *Science.* 2013; 342(6156): 373–77.

142 **decreased white blood cell count:** Zager A, Andersen ML, Ruiz FS, Antunes IB, Tufik S. Effects of acute and chronic sleep loss on immune modulation of rats. *American Journal of Physiology: Regulatory, Integrative and Comparative Physiology.* 2007; 293(1): R504–9.

142 **associated with increased tumor growth rate:** Everson CA. Functional consequences of sustained sleep deprivation in the rat. *Behavioral Brain Research.* 1995; 69(1–2): 43–54.

142 **growth hormone secretion in men:** Van Cauter E, Leproult R, Plat L. Age-related changes in slow-wave sleep and REM sleep and relationship with growth hormone and cortisol levels in healthy men. *Journal of the American Medical Association.* 2000; *284(7):* 861–68.

142 **working memory by as much as 38 percent:** Turner TH, Drummond SP, Salamat JS, Brown GG. Effects of 42-hr sleep deprivation on component processes of verbal working memory. *Neuropsychology.* 2007; *21*(6): 787–95.

143 **encourages the growth of dendritic protrusions:** Yang G, Lai CS, Cichon J, et al. Sleep promotes branch-specific formation of dendritic spines after learning. *Science.* 2014; *344*(6188): 1173.

143 **then taking three twenty-minute naps during the day:** Ray Williams Associates. Why the 8 hour sleep is a myth. June 12, 2014. http://ray williams.ca/blogs/why-the-8-hour-sleep-is-a-myth/?utm_source=rss& utm_medium=rss&utm_campaign=why-the-8-hour-sleep-is-a-myth

145 **body as occurring on the right:** Bareham CA, Manly T, Pustovaya OV, Scott SK, Bekinschtein TA. Losing the left side of the world: Rightward shift in human spatial attention with sleep onset. *Scientific Reports.* 2014; *4*: 5092.

145 **performance on a creative thinking test:** Home JA. Sleep loss and "divergent" thinking ability. *Sleep: Journal of Sleep Research & Sleep Medicine.* 1988; *11*(6): 528–36.

145 **the problem before taking a snooze:** Wagner U, Gals S, Halder H, Verleger R, Born J. Sleep inspires insight. *Nature.* 2004; *427*: 352–55.

145 **brain when processing the same information:** Walker MP, Liston C, Hobson JA, Stickgold R. Cognitive flexibility across the sleep-wake cycle: REM-sleep enhancement of anagram problem solving. *Cognitive Brain Research.* 2002; *14*: 317–24.

149 **causing it to use more oxygen than it does when it's awake:** Saladin KS. *Anatomy and Physiology: The Unity of Form and Function, 6th ed.* New York: McGraw-Hill; 2012: 537.

150 **the minds of the masses:** Brogaard B. The mad neuroscience of *Inception*. In: Botz-Bornstein T, ed. *Inception and Philosophy*. Chicago: Open Court; 2011.

151 **higher activity level of the frontal areas of the brain:** Voss U, Holzmann R, Hobson A, et al. Induction of self-awareness in dreams through frontal low current stimulation of gamma activity. *Nature Neuroscience*. 2014; *17*: 810–12.

152 **they are wearing something entirely different:** Brogaard B. The mad neuroscience of *Inception*. In: Botz-Bornstein T, ed. *Inception and Philosophy*. Chicago: Open Court; 2011.

153 **tends to prevent you from reaching a meaningful level of sleep:** Lee-chiong T. *Sleep Medicine: Essentials and Review*. New York: Oxford University Press; 2008: 52.

153 **sleepers have something in common with one another:** Peuhkuri K, Sihvola N, Korpela R. Diet promotes sleep duration and quality. *Nutrition Research*. 2012; 32(5): 309–19.

154 **so long as you use at least one bedsheet:** Onen SH, Onen F, Bailly D, Parquet P. Prevention and treatment of sleeping disorders through regulation of sleeping habits. *La Presse Médicale*. 1994; 23(10): 485–89.

154 **affect the timing of sleep cycles:** http://www.ncbi.nlm.nih.gov/pubmed/22889464

154 **the cushiness of the pillows:** A look inside the hotel room of the future. *The Independent*. June 9, 2011. http://www.independent.co.uk/travel/news-and-advice/a-look-inside-the-hotel-room-of-the-future-2295138.html

158 **a Grateful Dead concert in Oakland:** http://www.dreaminglucid.com/files/durso.pdf

Chapter 8: Super-Perceivers

166 **there's nothing it's like to be a bat:** Nagel T. What is it like to be a bat? *Philosophical Review*. 1974; 83(4): 435–50.

168 **visual image organized and in focus:** Eye scanpaths during visual imagery reenact those of perception of the same visual scene.

168 **these types of stimuli unconsciously:** Schwitzgebel E, Gordon MS. How well do we know our own conscious experience? The case of human echolocation. *Philosophical Topics.* 2000; *28*: 235–46.

169 **brain seeks to match up auditory and visual sensory inputs:** Vetter P, Smith FW, Muckli L. Decoding sound and imagery in early visual cortex. *Current Biology.* 2014; *24*: 1–7.

169 **detect objects in the environment through sound:** Kellogg W. Sonar system of the blind: New research measures their accuracy in detecting the texture, size, and distance of objects "by ear." *Science.* 1962; *137*(3528): 399–404.

169 **distance, size, shape, substance, and relative motion:** Stroffregen T, Pittenger J. Human echolocation as a basic form of perception and action. *Ecological Psychology.* 1995; *7*(3): 181–216.

169 **better able to echolocate if they are allowed to move:** Rosenblum LD, Gordon MS, Jarquin L. Echolocating distance by moving and stationary listeners. *Ecological Psychology.* 2000; *12*(3): 181–206.

169 **nonvisual sensory modalities like sound or touch:** Voss P, Lassonde M, Gougoux F, et al. Early- and late-onset blind individuals show supranormal auditory abilities in far-space. *Current Biology.* 2004; *14*(19); 1734–38.

169 **activation in the visual cortex in blind people:** Kujala T, Palva MJ, Solonen O, et al. The role of blind humans' visual cortex in auditory change detection. *Neuroscience Letters.* 2005; *379*(2): 127–31.

170 **after only a few hours of being blindfolded:** Pascual-Leone A, Amedi A, Fregni F, Merabet L. The plastic human brain cortex. *Annual Review of Neuroscience.* 2005; *28*(1): 377–401.

170 **somatosensory and auditory cortices were recruited less:** Sadato N, Pascual-Leone A, Grafman J, et al. Activation of the primary visual cortex by Braille reading in blind subjects. *Nature.* 1996; *380*: 526–28.

170 **simply activating the right neural structures after sensory deprivation:** Steves A, Weaver K. Functional characteristics of auditory cortex in the blind. *Behavioral Brain Research.* 2009; *196*(1): 134–38.

171 **sought to answer these questions in the 1940s:** Cotzin M, Dallenbach KM. "Facial vision": the role of pitch and loudness in the perception of obstacles by the blind. *American Journal of Psychology.* 1950; *63*(4): 485–515.

171 **sounds that was not available to the sighted controls:** Thaler L, Arnott S, Goodale M. Neural correlates of natural human echolocation in early and late blind echolocation experts. *PLOS One.* 2011; *6*(5): e20162.

172 **delays of approximately 10–12 nanoseconds:** Downey G. Getting around by sound: Human echolocation. *PLOS Blogs.* June 14, 2011. http://blogs .plos.org/neuroanthropology/2011/06/14/getting-around-by-sound -human-echolocation.

172 **leaving the mechanism in need of explanation:** Camhi JM. *Neuroethology.* Sunderland, MA: Sinauer; 1984.

178 **This condition is known as blindsight:** Weiskrantz L. *Blindsight: A Case Study and Implications.* Oxford: Oxford University Press; 1998.

179 **there are also reported cases of deaf hearing:** Garde M, Cowey A. "Deaf hearing": Unacknowledged detection of auditory stimuli in a patient with cerebral deafness. *Cortex.* 2000; *36*(1): 71–79.

181 **not normally associated with conscious experience:** Brogaard B. Conscious vision for action vs. unconscious vision for action. *Cognitive Science.* 2011; *35*: 1076–104. Brogaard B. Vision for action and the contents of perception. *Journal of Philosophy.* 2012; *109*(10): 569–87.

187 **the auditory cortex of congenitally deaf cats:** Meredith MA, Lomber SG. Somatosensory and visual crossmodal plasticity in the anterior auditory field of early-deaf cats. *Hearing Research.* 2011; *280*(1-2): 38–47.

187 **respond to peripheral visual stimulation:** http://www.nature.com /neuro/journal/v13/n11/full/nn.2653.html

188 **another study confirming similar results:** Karns CM, Dow MW, Neville HJ. Altered cross-modal processing in the primary auditory cortex of congenitally deaf adults: a visual-somatosensory fMRI study with a double-flash illusion. *Journal of Neuroscience.* 2012; 32(28): 9626–38.

188 **headphones during a brain imaging study:** http://www.ncbi.nlm.nih .gov/pubmed/22787048

191 **their names in the first couple of years of life:** Gardner C. The ears have it. *BBC Music Magazine.* 2009. http://perfectpitch.ucsf.edu/images /BBCMusicarticle.pdf

192 **perfect pitch after that critical period has closed:** Gervain J, Vines BW, Chen LM, et al. Valproate reopens critical-period learning of absolute pitch. *Frontiers in Systems Neuroscience.* 2013; 7: 102.

194 **a strong feeling of sadness, such as a crying sound effect:** Several emotionally painful examples of sadness, anger, and so forth are available on this site: http://www.soundsnap.com/tags/sad

195 **perfect pitch process musical notes differently:** Gardner, C. The ears have it. *BBC Music Magazine.* 2009. http://perfectpitch.ucsf.edu /images/BBCMusicarticle.pdf

Chapter 9: Brain Tech

198 **he developed an interest in music:** http://denver.cbslocal.com/2013 /11/20/teen-credits-concussion-with-giving-him-his-musical-talents

198 **more than 20 percent of scientists have used brain-enhancing drugs at some time:** Sahakian B, Morein-Zamir S. Professor's little helper. *Nature.* 2007; 450: 157–59.

199 **And both have rebound effects:** Nehlig A, Daval JL, Debry G. Caffeine and the central nervous system: mechanisms of action, biochemical, metabolic and psychostimulant effects. *Brain Research Reviews.* 1992; 17(2): 139–70. Westfall DP, Westfall TC. Miscellaneous sympathomimetic agonists. In: Brunton LL, Chabner BA, Knollmann BC, eds. *Goodman & Gilman's Pharmacological Basis of Therapeutics, 12th ed.* New York:

McGraw-Hill; 2010. Bolton S. Caffeine: psychological effects, use and abuse. *Orthomolecular Psychiatry.* 1981; *10*(3): 202–11.

200 **you can quickly complete the tasks on your plate:** Milgram NM, Callahan H, Siwak C. Adrafinil: a novel vigilance promoting agent. *CNS Drug Reviews.* 1999; 5: 193–212.

200 **hard hit on the head can have:** Brogaard B. Serotonergic hyperactivity as a potential factor in developmental, acquired and drug-induced synesthesia. *Frontiers in Human Neuroscience.* 2013; 7: 657.

203 **He wrote about his experiences for the *New York Times*:** Osborne L. Savant for a day. *New York Times Magazine.* June 22, 2003. http://www.nytimes.com/2003/06/22/magazine/22SAVANT.html

204 **almost all participants fail to solve it:** Chronicle EP, Ormerod JN, MacGregor JN. When insight just won't come: the failure of visual cues in the nine-dot problem. *Quarterly Journal of Experimental Psychology.* 2001; *54*: 903–19.

204 **even after a hundred attempts:** Weisberg RW, Alba JW. An examination of the alleged role of fixation in the solution of several insight problems. *Journal of Experimental Psychology: General.* 1981; *110*: 169–92.

205 **because our brains are structured in such a way as to limit creativity:** Kershaw TC, Ohlsson S. Multiple causes of difficulty in insight: the case of the nine-dot problem. *Journal of Experimental Psychology.* 2004; *30*: 3–13.

205 **our past experience when interpreting the world:** Snyder AW, Mitchell DJ. Is integer arithmetic fundamental to mental processing? The mind's secret arithmetic. *Proceedings of the Royal Society: Biology.* 1999; *266*: 597–92.

205 **but as an organized figure with strict boundaries:** Snyder AW, Bahramali H, Hawker T, Mitchell DJ. Savant-like numerosity skills revealed in normal people by magnetic pulses. *Perception.* 2006; 35: 837–45.

205 **This mimics the pattern of brain activity seen in many savants:**
Snyder AW. Explaining and inducing savant skills: privileged access
to lower level, less-processed information. *Philosophical Transactions of
the Royal Society: Biological Sciences.* 2009; *364:* 1399–405. Kaufman
SB. Conversations on creativity with Allan Snyder. *Psychology Today.*
January 13, 2010. http://www.psychologytoday.com/blog/beautiful-minds
/201001/conversations-creativity-allan-snyder

206 **while not affecting the motor performance of the dominant hand:**
Vines B, Cerruti C, Schlaug G. Dual-hemisphere tDCS facilitates greater
improvements for healthy subjects' nondominant hand compared to uni-
hemisphere stimulation. *BMC Neuroscience.* 2008; 9: 103.

207 **improve cognitive performance as well as general mood:** Barrett
DW, Gonzalez-Lima F. Transcranial infrared laser stimulation produces
beneficial cognitive and emotional effects in humans. *Neuroscience.* 2013;
230: 13–23.

207 **it just cares what kind of information it is:** O'Regan JK, Noe A. A
sensorimotor account of vision and visual consciousness. *Behavioral and
Brain Sciences.* 2001; *24*(5): 939–73.

207 **substitute touch for vision in blind individuals:** Bach-y-Rita P. Tactile sensory substitution studies. *Annals of New York Academic Sciences.*
2004; *1013*: 83–91.

208 **the region responsible for vision perception:** Bach-y-Rita P. *Brain
mechanisms in sensory substitution.* New York: Academic Press; 1972.

208 **one area for sensory experience in another:** Bach-y-Rita P. *Nonsyn-
aptic diffusion neurotransmission and late brain reorganization.* New
York: Demos-Vermande; 1995.

209 **Eagleman and Novich's recent success on Kickstarter:** https://
www.kickstarter.com/projects/324375300/vest-a-sensory-substitution
-neuroscience-project

212 **working on a game called *Meditation Deathmatch*:** http://meditation
deathmat.ch

213 **neurofeedback can help participants successfully manage pain:** deCharms RC, Christoff K, Glover GH, et al. Learned regulation of spatially localized brain activation using real-time fMRI. *NeuroImage.* 2004; *21*: 436–43.

213 **regulate emotion:** Hamilton JP, Glover GH, Hsu JJ, Johnson RF, Gotlib IH. Modulation of subgenual anterior cingulate cortex activity with real-time neurofeedback. *Human Brain Mapping.* 2001; *32*: 22–31.

213 **and improve working memory:** Zhang G, Yao L, Zhang H, Long Z, Zhao X. Improved working memory performance through self-regulation of dorsal lateral prefrontal cortex activation using real-time fMRI. *PLOS One.* 2013; *8*: e73735.

220 **It seems objectively unreasonable that it should, and yet it does:** Chalmers D. Facing up to the problem of consciousness. *Journal of Consciousness Studies.* 1995; *2*(3): 200–219.

220 **it should be able to do at a minimum:** Russell & Norvig 2003. http://www.amazon.com/Artificial-Intelligence-Modern-Approach-Edition/dp/0137903952

Luger & Stubblefield 2004. http://www.amazon.com/Algorithms-Data-Structures-Idioms-Prolog/dp/0136070477/ref=sr_1_2?s=books&ie=UTF8&qid=1413569934&sr=1-2

Poole, Mackworth & Goebel 1998. http://www.amazon.com/Computational-Intelligence-A-Logical-Approach/dp/0195102703

Nilsson 1998. http://books.google.com/books/about/Artificial_Intelligence.html?id=LIXBRwkibdEC

221 **as we know it, would cease to exist:** Ulam S. Tribute to John von Neumann. *Bulletin of the American Mathematical Society.* 1958; *64*(3): 1–49.

221 **these intelligent machines might take:** Vinge V. The coming technological singularity: how to survive in the post-human era. *Vision-21: Interdisciplinary Science and Engineering in the Era of Cyberspace.* NASA conference publication 10129; 1993. Kurzweil R. *The Singularity Is Near.* New York: Viking Books; 2005. Armstrong S, Sotala K. How

we're predicting AI—or failing to. In: Romportl J, Ircing P, Zackova E, Polak M, Schuster R, eds. *Beyond AI: Artificial Dreams*. Pilsen: University of West Bohemia; 2012: 52–75.

221 **appointed an intelligent robot to its board:** Zolfagharifard E. Would you take orders from a ROBOT? An artificial intelligence becomes the world's first company director. *MailOnline*. May 19, 2014. http://www .dailymail.co.uk/sciencetech/article-2632920/Would-orders-ROBOT -Artificial-intelligence-world-s-company-director-Japan.html

222 **navigation, and twerking robots:** http://www.bloomberg.com/video /meet-fonzie-the-dancing-robot-fCu_9N8AQhafapPg43F4DQ.html

Chapter 10: The Transcendent Everyman

226 **phonemic awareness in reading:** Massachusetts Institute of Technology. Brain scans may help diagnose dyslexia. *ScienceDaily*. August 13, 2013. www.sciencedaily.com/releases/2013/08/130813201424.htm

226 **people with dyslexia may be the result of differences in reading experience:** Krafnick AJ, Flowers DL, Luetje MM, Napoliello EM, Eden GF. An investigation into the origin of anatomical differences in dyslexia. *Journal of Neuroscience*. 2014; *34*(3): 901.

227 **involved in processing physical size:** Cohen Kadosh R, Cohen Kadosh K, Schuhmann T, et al. Virtual dyscalculia induced by parietal-lobe TMS impairs automatic magnitude processing. *Current Biology*. 2007; *17*: 689–93.

227 **equivalent to 3.6 years of natural brain aging:** http://stroke.ahajournals .org/content/37/1/263.full

228 **left to the right hemisphere during recovery:** Loring DW, Meador KJ, Lee GP, et al. Cerebral language lateralization: evidence from intra-carotid amobarbital testing. *Neuropsychologia*. 1990; *28*: 831–38.

229 **being conducted, testing this approach:** Study of attention deficit/ hyperactivity disorder using transcranial magnetic stimulation. http:// clinicaltrials.gov/show/NCT00001915

229 **processes math in the left hemisphere:** Dehaene S, Tzourio N, Frak V, et al. Cerebral activations during number multiplication and comparison: A PET study. *Neuropsychologia*. 1996; *34*: 1097–106.

Krueger F, Spampinato MV, Pardini M, et al. Integral calculus problem solving: an fMRI investigation. *NeuroReport*. 2008; *19*: 1095–99.

230 **the formation of novel visual imagery:** Yomogida Y, Sugiura M, Watanabe J, et al. Mental visual synthesis is originated in the fronto-temporal network of the left hemisphere. *Cerebral Cortex*. 2004; *14*(12): 1376–83.

INDEX